HOMOGENEOUS CATALYSIS

HOMOGENEOUS CATALYSIS
Mechanisms and Industrial Applications

SUMIT BHADURI
Department of Chemistry
Northwestern University
Evanston, Illinois

DOBLE MUKESH
ICI India R & T Centre
Thane, India

A John Wiley & Sons Publication.

New York • Chichester • Weinheim • Brisbane • Singapore • Toronto

Chemistry Library

Library of Congress Cataloging-in-Publication Data

Bhaduri, Sumit, 1948–
    Homogeneous catalysis : mechanisms and industrial applications / Sumit Bhaduri, Doble Mukesh.
        p.   cm.
    Includes bibliographical references and index.
    ISBN 0-471-37221-8 (acid-free paper)
    1. Catalysis.   2. Catalysis—Industrial applications.   I. Mukesh, Doble.   II. Title.
    TP156.C35.B52   2000
    660'.2995—dc21

                                                                99-045532

Printed in the United States of America

10 9 8 7 6 5 4 3 2 1

For
Vrinda Nabar—my severest critic and best friend
—Sumit Bhaduri

and

My dear wife Geetha
—Doble Mukesh

CONTENTS

PREFACE

This book has grown out of a graduate-level course on homogeneous catalysis that one of us taught at Northwestern University several times in the recent past. It deals with an interdisciplinary area of chemistry that offers challenging research problems. Industrial applications of homogeneous catalysis are proven, and a much wider application in the future is anticipated. Numerous publications and patent applications testify to the fact that in both the academic and industrial research laboratories the growth in research activity in this area in the past decade or so has been phenomenal.

Written mainly from a pedagogical point of view, this book is not comprehensive but selective. The material presented was selected on the basis of two criteria. We have tried to include most of the homogeneous catalytic reactions with proven industrial applications and well-established mechanisms. The basic aim has been to highlight the connections that exist between imaginative academic research and successful technology. In the process, topics and reports whose application or mechanism appears a little far-fetched at this point, have been given lower priority.

A chapter on the basic chemical concepts (Chapter 2) is meant for readers who do not have a strong background in organometallic chemistry. A chapter on chemical engineering fundamentals (Chapter 3) is included to give non-chemical engineering students some idea of the issues that are important for successful technology development. Because of the industrial mergers, acquisitions, etc., that have taken place over the past 10 years or so, the present names of some of the chemical companies today differ from their names as given in this book.

We have covered the literature up to the start of 1999. Recent publications that are particularly instructive or that deal with novel concepts are referred to in the answers to problems given at the end of each chapter. The sources for the material presented are listed in the bibliography at the end of each chapter.

Many people have helped in various ways in the preparation of this book: Professor James A. Ibers; Professor Robert Rosenberg and Virginia Rosenberg; Professor Du Shriver; Suranjana Nabar-Bhaduri and Vrinda Nabar; R. Y. Nadkar and V. S. Joshi. Sumit Bhaduri gratefully acknowledges a sabbatical leave from Reliance Industries Limited, India, without which the book could not have been completed. More than anything else, it was the students at Northwestern University whose enthusiastic responses in the classroom made the whole enterprise seem necessary and worthwhile. The responsibility for any shortcomings in the book is of course only ours.

SUMIT BHADURI

DOBLE MUKESH

CHAPTER 1

CHEMICAL INDUSTRY AND HOMOGENEOUS CATALYSIS

1.1 FEED STOCKS AND DEFINITIONS

Most carbon-containing feed stock is actually used for energy production, and only a very small fraction goes into making chemicals. The four different types of feedstock available for energy production are crude oil, other oils that are difficult to process, coal, and natural gas. Currently, the raw material for most chemicals is crude oil. Since petroleum is also obtained from crude oil, the industry is called petrochemical industry. Of the total amount of available crude oil, about 90% are sold as fuels of various kinds by the petroleum industry. It is also possible to convert sources of carbon into a mixture of carbon monoxide and hydrogen ($CO + H_2$), commonly known as synthesis gas. Hydrogen by itself is a very important raw material (e.g.,in the manufacture of ammonia). It is also required for the dehydrosulfurization of crude oil, a prerequisite for many other catalytic processing steps.

In this book we deal exclusively with *homogeneous* catalytic processes, that is, processes in which all the reactants are very often in gas–solution equilibrium. In other words,the catalyst and all the other reactants are in solution, and the catalytic reaction takes place in the liquid phase. In terms of total tonnage and dollar value, the contribution of homogeneous catalytic processes in the chemical industry is significantly smaller than that of *heterogeneous* catalytic reactions. All the basic raw materials or building blocks for chemicals are manufactured by a small but very important set of heterogeneous catalytic reactions. In these reactions gaseous reactants are passed over a solid catalyst. There are other reactions where liquid reactants are used with insoluble solid catalysts. These are also classified as heterogeneous catalytic reactions. Thus

1

in homogeneous catalytic reactions molecules of all the reactants, including those of the catalyst, are in the liquid phase. In contrast, in heterogeneous catalytic processes the molecules of the gaseous or liquid reactants are adsorbed on the surfaces of the solid catalysts. Unlike the discrete molecular structure of a homogeneous catalyst, a solid surface consists of an infinite array of ions or atoms.

1.2   FEED STOCK TO BASIC BUILDING BLOCKS BY HETEROGENEOUS CATALYSIS

To put the importance of homogeneous catalysis in perspective, we first present a very brief summary of the heterogeneous catalytic processes that are used to convert crude oil into the basic building blocks for chemicals. The heterogeneous catalytic reactions to which the feed stock is subjected, and the basic building blocks for chemicals that are obtained from such treatments, are shown in Fig. 1.1.

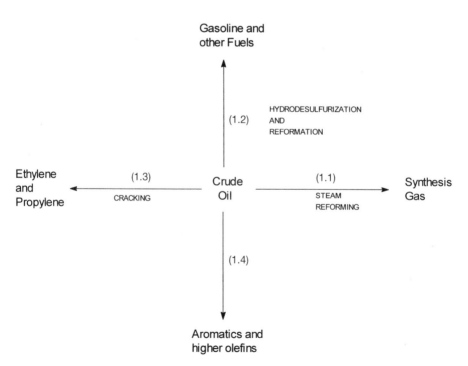

Figure 1.1   The basic building blocks for chemicals that are obtained from heterogeneous catalytic (and noncatalytic) treatment of crude petroleum.

Reaction 1.1 is known as steam reforming. The reaction conditions are fairly severe (>1000°C),and the structural strength of the catalyst is an important point of consideration. The catalyst employed is nickel on alumina, or magnesia, or a mixture of them. Other non-transition metal oxides such as CaO, SiO_2, and K_2O are also added.

Catalytic steam reforming could also be performed on natural gas (mainly methane) or the heavy fraction of crude oil called naphtha or fuel oil. The old method of producing synthesis gas by passing steam over red-hot coke was noncatalytic. Depending on the requirement for hydrogen, synthesis gas could be further enriched in hydrogen by the following reaction:

$$CO + H_2O \rightarrow CO_2 + H_2 \tag{1.5}$$

This is called the water gas shift reaction. We discuss this reaction in some detail in Chapter 4 (see Section 4.3). The heterogeneous catalysts used for the water gas shift reaction are of two types. The high-temperature shift catalyst is a mixture of Fe_3O_4 and Cr_2O_3 and operates at about 500°C. The low-temperature shift catalyst contains copper and zinc oxide on alumina, operates at about 230°C, and is more widely used in industry.

Step 1.2 involves separation of crude oil into volatile (<670°C) and non-volatile fractions. On fractional distillation, the volatile part gives hydrocarbons containing four or fewer carbon atoms, light gasoline, naphtha, kerosene, etc. All these could be used as fuels for different purposes. From the point of view of catalysis, the modification of the heavier fractions to "high octane" gasoline is important.

The conversion of the heavier fractions into high-octane gasoline involves two catalytic steps: the reduction of the level of sulfur in the heavy oil by hydrodesulfurization, followed by "reformation" of the hydrocarbon mixture to make it rich in aromatics and branched alkanes. Hydrodesulfurization prevents poisoning of the catalyst in the reformation reaction, and employs alumina-supported cobalt molybdenum sulfide. In this reaction sulfur-containing organic compounds react with added hydrogen to give hydrogen sulfide and hydrocarbons. The reformation reaction also requires hydrogen as a co-reactant and is carried out at about 450°C. The reformation reaction involves the use of acidified alumina-supported platinum and rhenium as the catalyst.

Reaction 1.3 is often called a *cracking reaction* because high-molecular-weight hydrocarbons are broken into smaller fragments. The major processes used for cracking naphtha into ethylene and propylene are noncatalytic and thermal, and are carried out at a temperature of about 800°C. However, there are other cracking reactions that involve the use of acidic catalysts, such as rare earth exchanged zeolites or amorphous aluminosilicates, etc. In some cracking reactions hydrogen is also used as a co-reactant, and the reaction is then called a *hydrocracking reaction*. Step 1.4 may involve all the catalytic and noncatalytic processes discussed so far.

1.3 BASIC BUILDING BLOCKS TO DOWNSTREAM PRODUCTS BY HOMOGENEOUS CATALYSIS

Although the fundamental processes for refining petroleum and its conversion to basic building blocks are based on heterogeneous catalysts, many important value-added products are manufactured by homogeneous catalytic processes. Some of these reactions are shown in Fig. 1.2.

The substances within the circles are the basic building blocks obtained from petroleum refining by processes discussed in the previous section. The products within the square are manufactured from these raw materials by homogeneous catalytic pathways. Except for 1.7, all the other four processes shown in Fig. 1.2 are large-tonnage manufacturing operations.

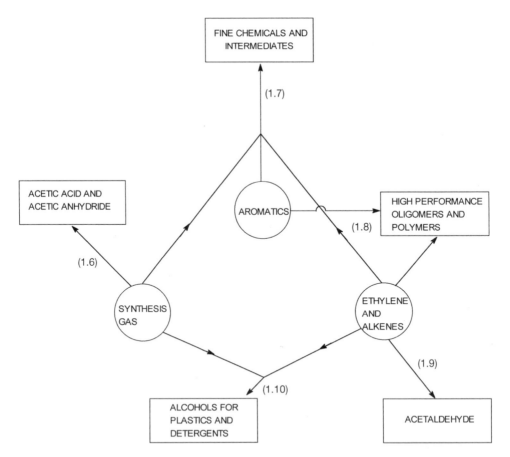

Figure 1.2 A few illustrative examples of chemicals and classes of chemicals that are manufactured by homogeneous catalytic processes. In 1.6 low-pressure methanol synthesis by a heterogeneous catalyst is one of the steps. In 1.9 it is ethylene that is converted to acetaldehyde. In 1.7 all the available building blocks may be used.

Step 1.6 involves the conversion of synthesis gas into methanol by a heterogeneous catalytic process. This is then followed by homogeneous catalytic carbonylation of methanol to give acetic acid. Similar carbonylation of methyl acetate gives acetic anhydride. These reactions are discussed in Chapter 4. Step 1.10 involves the conversion of alkenes and synthesis gas to aldehydes, which are then hydrogenated to give alcohols. These alcohols are used in plastics and detergents. The conversion of alkenes and synthesis gas to aldehydes is called an oxo or hydroformylation reaction and is discussed in Chapter 5. Step 1.9 is one of the early homogeneous catalytic processes and is discussed in Chapter 8. Steps 1.7 and 1.8 both represent the emerging frontiers of chemical technologies based on homogeneous catalysis. The use of metallocene catalysts in step 1.8 is discussed in Chapter 6.

As indicated by step 1.7, there are a number of small-volume but value-added fine chemicals, intermediates, and pharmaceuticals, where homogeneous catalytic reactions play a very important role. Some of these products, listed in Table 1.1, are optically active, and for these homogeneous catalysts exhibit almost enzymelike stereoselectivities. Asymmetric or stereoselective homogeneous catalytic reactions are discussed in Chapter 9.

1.4 COMPARISON AMONG DIFFERENT TYPES OF CATALYSIS

Heterogeneous catalysts are more widely used in industry than homogeneous catalysts because of their wider scope and higher thermal stability. There are no homogeneous catalysts as yet for cracking, reformation, ammonia synthesis, etc. The boiling point of the solvent and the intrinsic thermal stability of the catalyst also limit the highest temperature at which a homogeneous catalyst may be used. The upper temperature limit of a homogeneous catalytic reaction is about 250°C, while heterogeneous catalysts routinely operate at higher temperatures.

The two most important characteristics of a catalyst are its activity, expressed in terms of turnover number or frequency, and selectivity. The turnover number is the number of product molecules produced per molecule of the catalyst. The turnover frequency is the turnover number per unit time. In general, homogeneous and heterogeneous catalysts do not differ by an order of magnitude in their activities when either type of catalyst can catalyze a given reaction.

Selectivity could be of different type—chemoselectivity, regioselectivity, enantioselectivity, etc. Reactions 1.11–1.13 are representative examples of such selectivities taken from homogeneous catalytic processes. In all these reactions, the possibility of forming more than one product exists. In reaction 1.11 a mixture of normal and isobutyraldehyde rather than propane, the hydrogenation product from propylene, is formed. This is an example of chemoselectivity. Furthermore, under optimal conditions normal butyraldehyde may be obtained with more than 95% selectivity. This is an example of regioselectivity. Similarly, in reaction 1.12 the alkene rather than the alcohol functionality of allyl

TABLE 1.1 Products of Homogeneous Catalytic Reactions

Structure	Name and use	Process
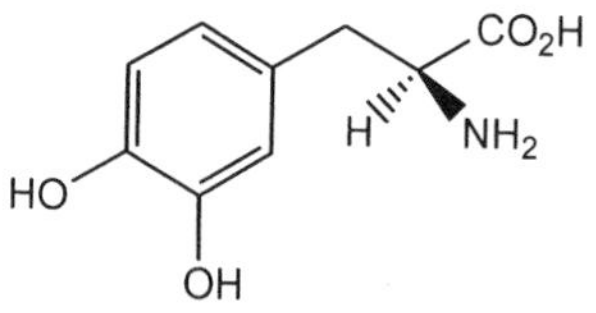	L-Dopa Drug for Parkinson's disease	Asymmetric hydrogenation
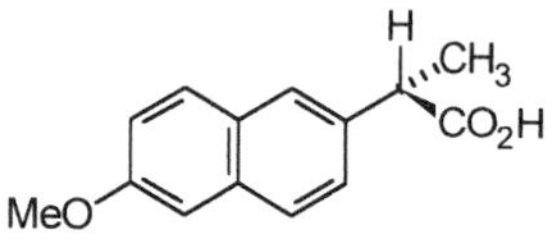 R - Liver toxin S - Anti-inflammatory	Naproxen® Anti inflammatory drug	Asymmetric hydroformylation or hydrocyanation or hydrogenation!
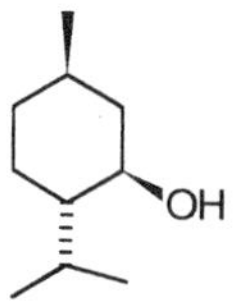	L-Menthol Flavoring agent	Asymmetric isomerization
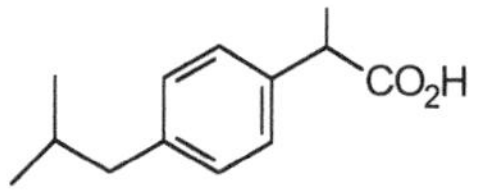	Ibuprofen Analgesic	Catalytic carbonylation
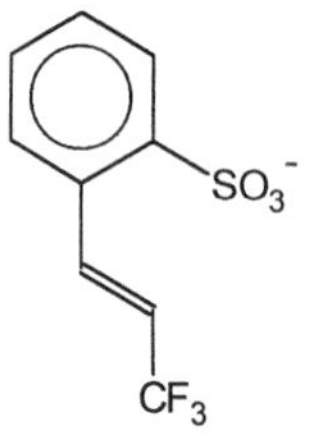	An intermediate for Prosulfuron Herbicide	C–C Coupling (Heck reaction)
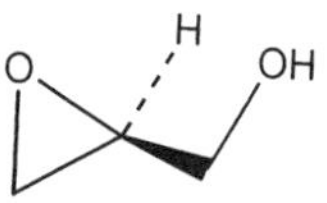	R-Glycidol One of the components of a heart drug	Asymmetric epoxidation (Sharpless epoxidation)

alcohol is selectively oxidized. However, the product epoxide, called glycidol, is a mixture of two enantiomers. In reaction 1.13 only one enantiomer of glycidol is formed in high yield. This is an example of an enantioselective reaction. Generally, by a choice of optimal catalyst and process conditions, it is possible to obtain very high selectivity in homogeneous catalytic reactions. This is one of the main reasons for the commercial success of many homogeneous-catalyst-based industrial processes.

$$\text{(1.11)}$$

$$\text{(1.12)}$$

$$\text{(1.13)}$$

Another important aspect of any catalytic process is the ease with which the products could be separated from the catalyst. For heterogeneous catalysts this is not a problem, since a solid catalyst is easily separated from liquid products by filtration or decantation. In some of the homogeneous catalytic processes, catalyst recovery is a serious problem. This is particularly so when an expensive metal like rhodium or platinum is involved. In general, catalyst recovery in homogeneous catalytic processes requires careful consideration.

Finally, for an overall perspective on catalysis of all types, here are a few words about biochemical catalysts, namely, enzymes. In terms of activity, selectivity, and scope, enzymes score very high. A large number of reactions are catalyzed very efficiently, and the selectivity is high. For chiral products enzymes routinely give 100% enantioselectivity. However, large-scale application of enzyme catalysis in the near future is unlikely for many reasons. Isolation of a reasonable quantity of pure enzyme is often very difficult and expensive. Most enzymes are fragile and have poor thermal stability. Separation of the enzyme after the reaction is also a difficult problem. However, in the near future, catalytic processes based on thermostable enzymes may be adopted for selected products.

The above-mentioned factors—activity, selectivity, and catalyst recovery—are the ones on which comparison between homogeneous and heterogeneous catalysts is normally based. Other important issues are catalyst life, susceptibility towards poisoning, diffusion, and last but probably most important, control of performance through mechanistic understanding. The life of a homogeneous catalyst is usually shorter than that of a heterogeneous one. In practical terms this adds to the cost of homogeneous catalytic processes, since the metal has to be recovered and converted back to the active catalyst. Although homogeneous catalysts are thermally less stable than heterogeneous ones, they are less susceptible to poisoning by sulfur-containing compounds. Another important difference between the two types of catalysis is that macroscopic diffusion plays an important role in heterogeneous catalytic processes but is less important for homogeneous ones.

Finally, the biggest advantage of homogeneous catalysis is that, in most cases, the performance of the catalyst can be explained and understood at a molecular level. This is because the molecular species in a homogeneous catalytic system are easier to identify than in a heterogeneous one. For soluble catalysts, there are many relatively simple spectroscopic and other techniques for obtaining accurate information at a molecular level (see Section 2.5). In contrast, the techniques available for studying adsorbed molecules on solid surfaces are more complex, and the results are often less unequivocal. Based on a mechanistic understanding, the behavior of a homogeneous catalyst can be fine-tuned by optimal selection of the metal ion, ligand environment and process conditions. As an example we refer back to reaction 1.11. In the absence of any phosphorus ligand and relatively high pressures, the ratio of the linear to the branched isomer is about 1:1. However, by using a phosphorus ligand and lower pressure, this ratio could be changed to >19:1. This change in selectivity can be explained and in fact can be predicted on the basis of what is known at a molecular level.

To summarize, both heterogeneous and homogeneous catalysts play important roles in the chemical industry. Roughly 85% of all catalytic processes are based on heterogeneous catalysts, but homogeneous catalysts, owing to their high selectivity, are becoming increasingly important for the manufacture of tailor-made plastics, fine chemicals, pharmaceutical intermediates, etc.

1.5   WHAT IS TO FOLLOW—A SUMMARY

In the following chapters we discuss the mechanisms of selected homogeneous catalytic reactions. Brief descriptions of some of these reactions, the metals involved, and the chapters where they are to be found are given in Table 1.2. The following points deserve attention: First, the names of five reactions (see the second to the sixth row) begin with the prefix "hydro." In all these reactions a hydrogen atom and some other radical or group are added across the double bond of an alkene. Thus a "hydroformylation" reaction comprises an addition of H and CHO; "hydrocyanation," an addition of H and CN, etc. Second, from the fourth column it is clear that complexes of a variety of transition and occasionally other metals have been successfully used as homogeneous catalysts. Third, the last row includes most of the reactions of the previous rows with an important modification, namely, the use of chiral metal complexes as catalysts. In the next two chapters we discuss some fundamental chemical and engineering concepts of homogeneous catalysis. These concepts will help us to understand the behavior of different homogeneous catalytic systems and their successful industrial implementation.

TABLE 1.2 Important Homogeneous Catalytic Reactions

Common name	Reactant	Product	Metal
Carbonylation	1. Methanol and CO	1. Acetic acid	1. Rh or Co
	2. Methyl acetate and CO	2. Acetic anhydride	2. Rh
	3. Methyl acetylene, CO, methanol	3. Methyl methacrylate	3. Pd
Hydrocarboxylation	Alkene, water, and CO	Carboxylic acid	All in Chapter 4
Hydroformylation	1. Propylene, CO, H₂	1. *n*-Butyraldehyde	1. Rh or Co
	2. α-alkenes, CO, H₂	2. *n*-Aldehydes	2. Co
			Chapter 5
1. Hydrocyanation	1. Butadiene and HCN	1. Adiponitrile	1. Ni
2. Hydrosilylation	2. Alkene and R₃SiH	2. Tetraalkylsilane	2. Pt
3. Hydrogenation	3. Alkene or aldehyde and H₂	3. Alkane or alcohol	3. Rh or Co
4. Metathesis	4. Alkenes or dienes	4. Rearranged alkene(s) or dienes	4. Mo, Re, or Ru
			All in Chapter 7
Polymerization	Ethylene, propylene, etc.	Polymers	Ti or Zr with Al; also Cr
			Chapter 6
Di- and oligomerization	Propylene, ethylene, etc.	Oligomers	Ni
			Chapter 7
Auto-oxidation	Cyclohexane or *p*-xylene	Adipic or terephthalic acid	Co, Mn, V
Epoxidation	Propylene	Propylene oxide	Mo
Wacker reaction	Ethylene and O₂	Acetaldehyde	Pd and Cu
			All in Chapter 8
Asymmetric reactions	Mainly alkenes with other appropriate reactants	Chiral products of different kinds	Rh, Ru, Ir, Cu, Ti, Mn, Co, Os, La, etc. Chapter 9

9

PROBLEMS

1. In a hydrogenation reaction with a soluble catalyst there are liquid and gaseous phases present. Why is the reaction called homogeneous rather than heterogeneous?

Ans. Reaction takes place between dissolved gas, catalyst, and the substrate, that is, all in one phase with discrete molecular structures.

2. Write a hypothetical single-step catalytic route for all the compounds shown in Table 1.1.

Ans. Many possibilities, but for actual catalytic processes see Chapters 4, 6, and 9.

3. In Question 2 highlight the chemo-, regio-, and steroselectivity, if any, that is involved in the hypothetical routes.

Ans. All but ibuprofen and Prosulfuron are enantioselective. Prosulfuron and ibuprofen are chemoselective.

4. The chances of success are greater if one tries to develop a homogeneous water gas shift catalyst rather than a steam reformation catalyst. Why?

Ans. Thermodynamics highly unfavorable at temperatures at which a homogeneous catalyst is stable.

5. Propylene oxide (PO) used to be made by reacting propylene with chlorine and water (hypochlorous acid) to give chlorohydrin followed by its reaction with calcium hydroxide. Methyl methacrylate ($CH_3(CO_2CH_3)C{=}CH_2$) used to be made by the reaction of acetone with HCN followed by hydrolysis with sulfuric acid. The solid wastes generated in these two processes were $CaCl_2$ and NH_4HSO_4, respectively. Using the concept of atom utilization [atom utilization = $100 \times$ (mol. wt. of the desired product/total mol. wt. of all products)], show how homogeneous catalytic routes are superior.

Ans. In the homogeneous catalytic process for PO the by-product is *t*-butanol, which has an attractive market. The atom utilization by the old route for PO is 31%. The atom utilizations by the new route are 44 and 56% for PO and *t*-butanol. For methylmethacrylate the atom utilization by the new route (methyl acetylene plus carbon monoxide and methanol) is 100%, and by the old route is 46% (see R. A. Sheldon, *Chemtech*, 1994, March, 38–47).

6. One of the synthetic routes for the anticancer drug Taxol, which has twelve stereo centers, involves a homogeneous C–C coupling reaction. The industrial production of a protease inhibitor that has stereospecific arrangements of amino and hydroxyl groups on two adjacent carbon atoms also involves homogeneous catalysis. From Table 1.1 identify the possible reaction types that are used in these two syntheses.

Ans. Heck reaction and Sharpless epoxidation followed by opening of the epoxide with amine (see W. A. Herrmann et al., *Angew. Chem. Int. Ed.,* 1997, **36,** 1049–1067).

7. Industrial manufacturing processes for acrylic acid and acrylonitrile are based on selective oxidation and ammoxidation of propylene using heterogeneous catalysts. For acrylic acid a pilot-scale homogeneous catalytic route (Pd, Cu catalysts) involves ethylene, carbon monoxide, and oxygen as the starting materials. What are the factors that need to be taken into account before the homogeneous catalytic route may be considered to be a serious contender for the synthesis of acrylic acid?

Ans. All the factors listed in Section 1.4, especially catalyst separation. The relative cost and availability of ethylene and propylene also need to be considered. For a history of acrylic acid manufacturing routes, see the reference given in the answer to Problem 6.

BIBLIOGRAPHY

Sections 1.1 and 1.2

Books

Heterogeneous Catalysis: Principles and Applications, G. C. Bond, Clarendon Press, New York, 1987.
Catalytic Chemistry, B. C. Gates, Wiley, New York, 1991.
Principles and Practice of Heterogeneous Catalysis, J. M. Thomas and W. J. Thomas, VCH, New York, 1997.

Section 1.3

Books

Homogeneous Catalysis: The Applications and Chemistry of Catalysis by Soluble Transition Metal Complexes, G. W. Parshall and S. D. Ittel, Wiley, New York 1992.
Applied Homogeneous Catalysis with Organometallic Compounds, Vols. 1 & 2, edited by B. Cornils and W. A. Herrmann, VCH, Weinheim, New York, 1996.
Handbook of Co-Ordination Catalysis in Organic Chemistry, P. A. Chaloner, Butterworths, London, 1986.
Homogeneous Transition Metal Catalysis: A Gentle Art, C. Masters, Chapman and Hall, New York, 1981.
Homogeneous Catalysis with Metal Phosphine Complexes, edited by L. H. Pignolet, Plenum Press, New York, 1983.
Principles and Applications of Homogeneous Catalysis, A. Nakamura and M. Tsutsui, Wiley, New York, 1980.
Homogeneous Catalysis with Compounds of Rhodium and Iridium, R. S. Dickson, D. Reidel, Boston, 1995.

Articles

G. W. Parshall and W. A. Nugent, *Chemtech*, **18**(3), 184–90, 1988; *ibid.* **18**(5), 314–20, 1988; *ibid.* **18**(6), 376–83, 1988.

G. W. Parshall and R. E. Putscher, *J. Chem. Edu.*, **63**, 189–91, 1986.

Section 1.4

The texts and the articles given under Section 1.3. Also see the references given in answers to Problems 5 and 6.

CHAPTER 2

───

BASIC CHEMICAL CONCEPTS

───

In this chapter we discuss some of the basic concepts of organometallic chemistry and reaction kinetics that are of special relevance to homogeneous catalysis. The catalytic activity of a metal complex is influenced by the characteristics of the central metal ions and the attached ligands. We first discuss the relevant properties of the metal ion and then the properties of a few typical ligands.

2.1 THE METAL

Insofar as the catalytic potential of a metal complex is concerned, the formal charge on the metal atom and its ability to form a bond of optimum strength with the incoming substrate are obviously important. We first discuss a way of assessing the charge and the electronic environment around the metal ion. The latter is gauged by the "electron count" of the valence shell of the metal ion.

2.1.1 Oxidation State and Electron Count

The formal charge assigned to a metal atom in a metal complex is its oxidation state. The sign of the charge for metal is usually positive, but not always. It is assigned and justified on the basis of relative electronegativities of the central metal atom and the surrounding ligands. The important point to note is that a fully ionic model is implicit, and to that extent the formal oxidation state may not correspond to the real situation. It does not take into account the contribution from covalency, that is, electrons being shared between the metal atom and the ligand, rather than being localized either on the ligand or on the metal.

A few examples of special relevance to homogeneous catalytic systems are given in Fig. 2.1, along with total electron counts. The rationales behind the schemes that are used to arrive at the electron counts are described in the following.

Electron counting could be done either after assignment of an oxidation state to the metal (i.e., assuming ionic character in the bonds) or without assigning any oxidation state (i.e., assuming full covalency and zero oxidation state of the metal). In the latter case, the counting is very similar to the procedure of counting electrons in CH_4, NH_3, etc. to arrive at the octet rule. Both ways of counting electrons are illustrated.

RhCl(PPh$_3$)$_3$: The chlorine radical (Cl˙) accepts an electron from rhodium metal (electronic configuration $4d^7,5s^2$) to give Cl^- and Rh^+. The chloride ion then donates two electrons to the rhodium ion to form a dative or a coordinate bond. Each PPh$_3$ donates a lone pair of electrons on the phosphorus atom to the rhodium ion. The total number of electrons around rhodium is therefore 8 + 2 + 3 × 2 = 16, and the oxidation state of rhodium is obviously 1+. The other way of counting is to take the nine electrons of rhodium and add one electron for the chlorine radical and six for the three neutral phosphine ligands. This also gives the same electron count of 16.

	Oxidation State	Electron count
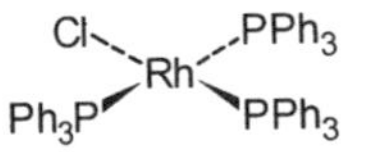 Cl—Rh(PPh₃)(PPh₃)(Ph₃P)	1+	16
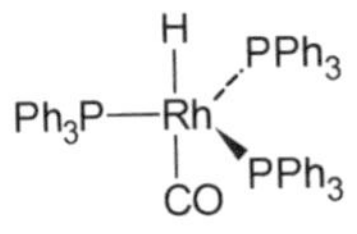 Ph₃P—Rh(H)(PPh₃)(CO)	1+	18
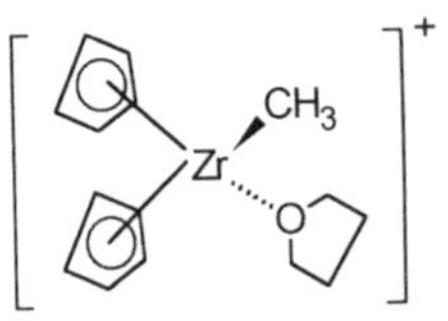 [Zr(Cp)₂(CH₃)(O-THF)]⁺	4+	16
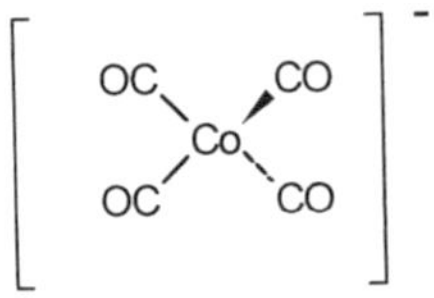 [Co(CO)₄]⁻	1-	18

Figure 2.1   Formal oxidation states and valence electron counts of metal ions in some homogeneous catalysts.

Similarly, for RhH(CO)(PPh$_3$)$_3$ the rhodium oxidation state is 1+ because the hydrogen atom is assumed to carry, with some justification, a formal negative charge. The five ligands, H$^-$, CO, and three PPh$_3$, each donate two electrons, and the electron count therefore is $8 + 5 \times 2 = 18$. With the covalent model the hydrogen ligand is treated as a radical, rhodium is considered to be in a zero oxidation state, and the electron count is $9 + 1 + 4 \times 2 = 18$.

[Cp$_2$Zr(CH$_3$)(THF)]$^+$: The zirconium oxidation state is 4+ and each Cp$^-$ ligand donates six electrons. The ligand CH$_3^-$ donates two electrons. The solvent molecule, THF, also donates two electrons, and the total electron count is $12 + 0 + 2 + 2 = 16$. With the covalent model zirconium is in the zero oxidation state and has four electrons ($4d^2,5s^2$) in the valence shell. Both Cp and CH$_3$ are considered as radicals and therefore donate five and one electron, respectively. The valence electron count is therefore $4 + 2 \times 5 + 1 + 2 - 1 = 16$. Notice that because of the positive charge, we subtract one electron.

Co(CO)$_4^-$: Since there is a net negative charge and CO is a neutral ligand, the formal oxidation state of cobalt is 1−. The electron count is therefore $10 + 4 \times 2 = 18$. According to the covalent model, the electron count is also $9 + 4 \times 2 + 1 = 18$, but cobalt is assumed to be in a zero oxidation state, and one electron is added for the negative charge.

It should be clear from the preceding examples that as long as we are consistent in our ways of counting electrons, either method will give the same the answer. Like the octet rule for the first-row elements, there is an 18-electron rule for the transition metals. The rationale behind this rule is simply that the metal ion can use nine orbitals—five d orbitals, three p orbitals, and one s orbital—for housing electrons in its valence shell. Methane, water, etc. are stable molecules, as they have eight electrons around the central atom. Similarly, organometallic complexes that have 18 electrons in the outer shell are stable complexes. This rule is often referred to as the "eighteen-electron rule" or the rule of effective atomic number (EAN).

2.1.2 Coordinative Unsaturation

Complexes that have CO, PPh$_3$, H$^-$ etc. as ligands tend to be reactive if the electron count is less than eighteen. They undergo reactions to form extra bonds so that an electron count of 18 is reached. When the electron count is less than 18, the metal complex is often classified as coordinatively unsaturated. Among the complexes shown in Fig. 2.1, RhCl(PPh$_3$)$_3$ is coordinatively unsaturated, while RhH(CO)(PPh$_3$)$_3$ and Co(CO)$_4^-$ are coordinatively saturated. Apart from RhCl(PPh$_3$)$_3$, there are many other reactive complexes with an electron count of 16.

High reactivity may also result from easy displacement of weakly bound ligands. The zirconium compound [Cp$_2$Zr(THF)(CH$_3$)]$^+$, shown in Fig. 2.1 and discussed in the previous section, is an example. Unlike RhCl(PPh$_3$)$_3$, which tries to form extra bonds in its reactions, the zirconium compound's reactivity in a catalytic reaction is due to the easy displacement of THF by the substrate.

The term *coordinative unsaturation* includes this type of reactivity also. In other words, the ability to form extra bonds or facile displacement of weakly bound ligands, which in many cases may just be solvent molecules, are both manifestations of coordinative unsaturation.

Coordinative unsaturation can sometimes be induced by using bulky ligands. A few such ligands can take up most of the space around the metal atom and prevent the presence of a full complement of ligands. So due to steric constraints, an 18-electron count, which only a full complement of ligands can give, is not achieved. As an example, nickel in the zero oxidation state requires the presence of four monodentate phosphorus ligands to give an electron count of 18. However, if these ligands are bulky, then steric repulsion between them causes ligand dissociation, and the following equilibrium is established. The species NiL_3 on the right-hand side of the equilibrium has an electron count of 16 and is coordinatively unsaturated.

$$NiL_4 \rightleftharpoons NiL_3 + L \qquad (2.1)$$

A quantitative estimation of the steric demand of L can be made in terms of its cone angle. As shown in Fig. 2.2, it is the angle of a cone with its vertex at the metal atom and a metal–phosphorus distance of about 22.8 nm. The cone is created by the surface that just encloses all the ligand atoms for all orientations resulting from the rotation around the metal–phosphorus bond.

Finally, many complexes that participate in homogeneous catalytic reactions have electron counts less than 16. This is especially true for high-oxidation-state early-transition-metal complexes such as $(C_2H_5)TiCl_3$, $Ti(OPr^i)_4$, etc. Catalytically active, late-transition-element complexes with electron counts less than sixteen are also known. An important example is $RhCl(PPh_3)_2$, a 14-electron complex that plays a crucial role in homogeneous hydrogenation reactions (see Section 7.3.1).

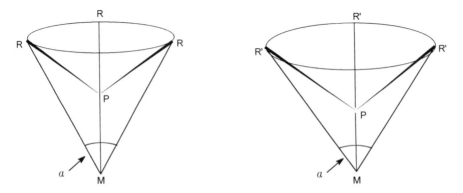

Figure 2.2   Cone angles (α) with two different phosphines PR_3 and PR_3'. In both the M–P distance is ~22.8 nm.

2.1.3 Rare Earth Metals

The examples discussed so far are all transition metal complexes. As we will see later (Chapters 4–9), most homogeneous catalytic processes are indeed based on transition metal compounds. However, catalytic applications of rare earth complexes have also been reported, although so far there has not been any industrial application. Of special importance are the laboratory-scale uses of lanthanide complexes in alkene polymerization and stereospecific C–C bond formation reactions (see Sections 6.4.3 and 9.5.4).

The points to note are that, unlike transition-metal-based homogeneous catalysts where the metal ions can have a wide range of oxidation states, the rare earth ions in almost all cases are in the $3+$ oxidation state. Also, electron count has little significance for these complexes. There is a similarity between high-oxidation-state early-transition-metal complexes and those of rare earths. In both the cases ligand dissociation or some other similar mechanism generates coordinative unsaturation. In both cases the substrates are activated by direct interaction with small, highly electropositive metal ions. Finally, in both cases the oxidation states of the metal ions do not change during catalysis.

2.2 IMPORTANT PROPERTIES OF LIGANDS

A very large number of different types of ligands can coordinate to transition metal ions. Once coordinated the reactivity of the ligands may dramatically change. Here we first discuss some of the ligands that are often involved in homogeneous catalytic reactions.

2.2.1 CO, $R_2C=CR_2$, PR_3, and H^- as Ligands

The traditional definition of a coordinate or a dative bond is the donation and sharing of electrons, usually a lone pair, by the ligand onto and with the metal. In other words, all ligands behave as Lewis bases, and the metal ion acts as a Lewis acid. The ligands listed are no exception insofar as the electron donation part is concerned. Both CO and PPh_3 donate lone pairs on the carbon and the phosphorus atoms, respectively. With alkenes, since there are no lone pairs, it is the π electrons that are donated and shared. Similarly, a gain of one electron from the metal by the hydrogen atom produces a hydride ligand, which then donates and shares the electron pair with the metal ion.

The first three of these ligands, and there are a few others, such as N_2, NO, etc., differ from H_2O, NH_3, etc. in that they also accept electron density from the metal; that is, they act as Lewis acids. The electron density is often accepted in an orbital of π^*, symmetry and in such cases the ligands are called π acid ligands. The donation of electron density by the metal atom to the ligand is also referred to as back-donation. While H^- does not act as a Lewis acid, dihydrogen does. In fact, such an interaction is responsible for the formation of stable metal–dihydrogen complexes. In the extreme case where two electrons are formally transferred back to dihydrogen from the metal, the H–H bond breaks, and two H^- ligands are formed. For this reason we consider H^- along with π acid ligands. Fig. 2.3 shows the Lewis acid–like behavior of CO, C_2H_4, and H_2 in terms of overlaps between empty ligand and filled metal orbitals of compatible symmetry. Back-donation is a bonding interaction between the metal atom and the ligand because the signs of the donating metal d orbitals and the ligand π^* (σ^* for H_2) acceptor orbitals match.

The π ligands play important roles in a large number of homogeneous catalytic processes. Alkene polymerization and a variety of other reactions involve alkene coordination (see Chapters 6 and 7). As the name suggests, CO is the main ligand in carbonylation reactions (see Chapter 4). All four ligands: CO, alkene, H^-, and PR_3, play important parts in hydroformylation reactions (see Chapter 5).

2.2.2 Alkyl, Allyl, and Alkylidene Ligands

Alkyl complexes are intermediates in a number of homogeneous catalytic processes, such as carbonylation, alkene polymerization, hydrogenation, etc. Allyl

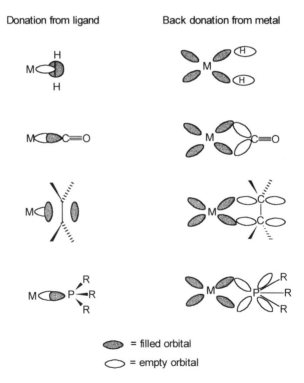

Figure 2.3 Schematic presentation of orbital overlaps for metal–ligand bond formations.

ligands are important in catalytic hydrocyanation reactions, and a number of other reactions where butadiene is used as one of the starting materials (Chapter 7). Alkylidene intermediates are involved in alkene metathesis reactions. Representative examples of these three ligands are given in Fig. 2.4.

All these ligands have extensive chemistry; here we note only a few points that are of interest from the point of view of catalysis. The relatively easy formation of metal alkyls by two reactions—insertion of an alkene into a metal–hydrogen or an existing metal–carbon bond, and by addition of alkyl halides to unsaturated metal centers—are of special importance. The reactivity of metal alkyls, especially their kinetic instability towards conversion to metal hydrides and alkenes by the so-called β-hydride elimination, plays a crucial role in catalytic alkene polymerization and isomerization reactions. These reactions are schematically shown in Fig. 2.5 and are discussed in greater detail in the next section.

Alkylidene complexes are of two types. The ones in which the metal is in a low oxidation state, like the chromium complex shown in Fig. 2.4, are often referred to as Fischer carbenes. The other type of alkylidene complexes has the metal ion in a high oxidation state. The tantalum complex is one such example. For both the types of alkylidene complexes direct experimental evidence of the presence of double bonds between the metal and the carbon atom comes from X-ray measurements. Alkylidene complexes are also formed by α-hydride elimination. An interaction between the metal and the α-hydrogen atom of the alkyl group that only weakens the C–H bond but does not break it completely is called an *agostic interaction* (see Fig. 2.5). An important reaction of alkylidene complexes with alkenes is the formation of a metallocycle.

Allyl ligands have features common to metal–alkyl and metal–olefin complexes, and can act as three-electron donor ligands according to the electron counting scheme where total covalency is assumed. The metal ion in these complexes interacts with all three carbon atoms of the allyl functionality in an equivalent manner.

2.3 IMPORTANT REACTION TYPES

Almost all homogeneous catalytic processes involve a relatively small set of typical reactions. We have already seen a few of these in the previous sections. Here we discuss them in greater detail and introduce a few others.

2.3.1 Oxidative Addition and Reductive Elimination

Oxidative addition is a reaction where the metal undergoes formal oxidation and atoms, groups of atoms, or molecules are added to the metal center. Reductive elimination is the exact opposite of oxidative addition—the metal ion is formally reduced with elimination of ligands. A few examples are shown in Fig. 2.6. In all these examples the forward reactions are oxidative addition

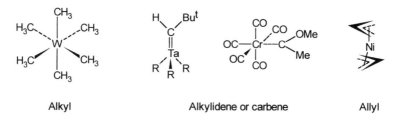

Alkyl   Alkylidene or carbene   Allyl

Figure 2.4 Typical examples of alkyl, alkylidene (carbene), and allyl complexes.

(increase in oxidation states by 2), and the reverse reactions are reductive elimination (decrease in oxidation states by 2).

   All the forward reactions are important steps in commercial homogeneous
catalytic processes. Reaction 2.2 is a step in methanol carbonylation (see Chapter 4), while reaction 2.3 is a step in the hydrogenation of an alkene with an
acetamido functional group. This reaction, as we will see in Chapter 9, is

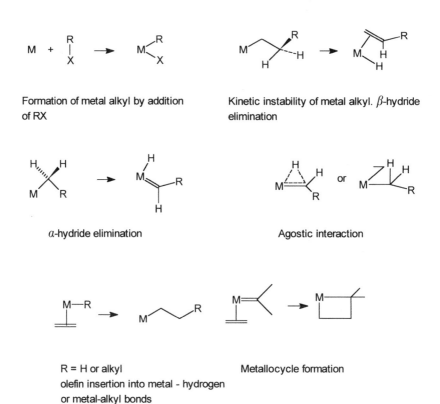

Formation of metal alkyl by addition
of RX

Kinetic instability of metal alkyl. β-hydride
elimination

α-hydride elimination

Agostic interaction

R = H or alkyl
olefin insertion into metal - hydrogen
or metal-alkyl bonds

Metallocycle formation

Figure 2.5 Some important reactions and interactions of metal–alkyl and metal–
alkylidene complexes.

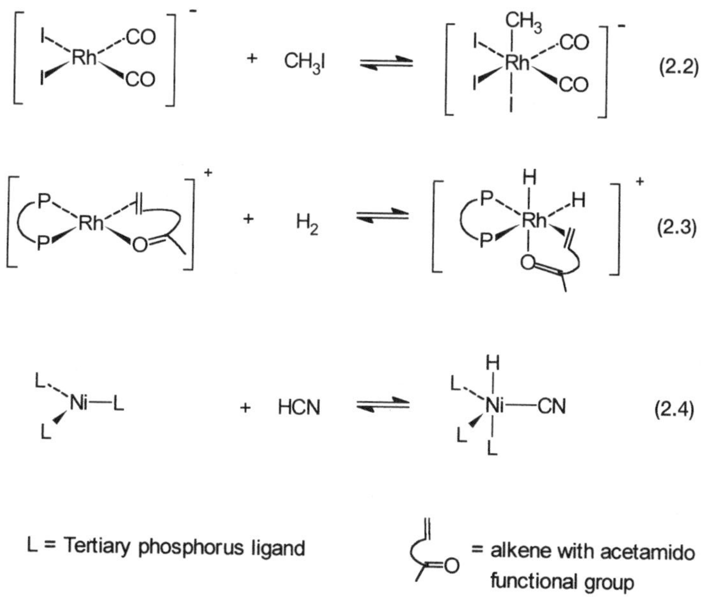
CH3
CO
Rh
CO
CH3I
(2.2)
H
P
Rh
H2
(2.3)
HCN
Ni
L
CN
(2.4)
L = Tertiary phosphorus ligand
= alkene with acetamido functional group

Figure 2.6 Representative examples of oxidative addition and reductive elimination reactions.

important for asymmetric hydrogenation. Reaction 2.4 is the first step in the hydrocyanation of butadiene for the manufacture of adiponitrile (Chapter 7).

As already mentioned, the reverse reactions of Fig. 2.6 are reductive elimination reactions. By the principle of microscopic reversibility, the existence of an oxidative addition reaction means that reductive elimination, if it were to take place, would follow the reverse pathway. The reductive elimination of an alkane from a metal-bonded alkyl and hydride ligand in most cases poses a mechanistic problem. This is because clean oxidative addition of an alkane onto a metal center to give a hydrido metal alkyl, such as a reaction like Reaction 2.5, is rare.

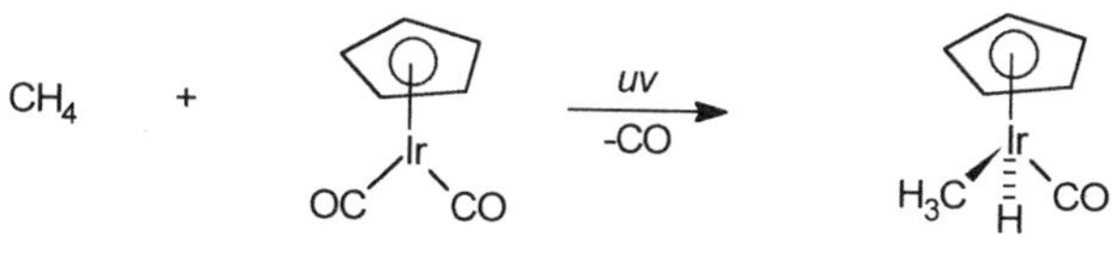
CH4
uv
-CO
Ir
OC
CO
H3C
H
CO

(2.5)

The mechanism of reductive elimination of a hydrido alkyl complex is therefore often approached in an indirect manner. The hydrido–alkyl complex is made not by oxidative addition of the alkane but by some other route. The decomposition of the hydrido–alkyl complex to give alkane is then studied for mechanistic information. Reductive eliminations of an aldehyde from an acyl–hydrido complex, Reaction 2.7, and acetyl iodide from an iodo–acyl complex,

Reaction 2.6, are important steps in catalytic hydroformylation and carbonylation reactions, respectively.

$$(2.6)$$

$$(2.7)$$

2.3.2   Insertion Reactions

In homogeneous catalytic reactions, old bonds are usually broken by oxidative addition reactions and new bonds are formed by reductive elimination and insertion reactions. A few representative examples that are of relevance to catalysis are shown by Reactions 2.8–2.11. The following points deserve attention. Reactions 2.8, 2.9, and 2.10 are crucial steps in hydrogenation, polymerization, and CO-involving catalytic reactions. Reaction 2.8 is, of course, just the reverse of β-hydride elimination. Sometimes this reaction is also called a "hydride attack" or "hydride transfer" reaction.

Insertion of olefin into M-H bond   (2.8)

Insertion of olefin into M-R bond   (2.9)

Insertion of CO into M-R bond   (2.10)

Insertion of CO into M-H bond   (2.11)

Calling Reaction 2.10 an insertion of CO into a metal–alkyl bond may be misleading. There is evidence to show that the alkyl group actually migrates to CO. A more appropriate description of a number of insertion reactions is "migratory insertion." However, in the rest of this book, we will ignore this mechanistic distinction and simply call Reactions 2.8–2.11 as insertion reactions.

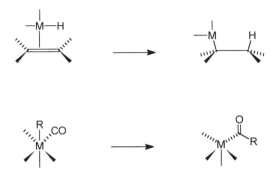

Figure 2.7 *cis* insertions of alkene and CO into metal–hydrogen and metal–alkyl bonds, respectively.

Reaction 2.11, although proposed in CO hydrogenation reactions, is thermodynamically unfavorable, that is, there is a net loss in bond energy in breaking a metal–hydrogen bond and forming a carbon–hydrogen one. There is so far no clear example of insertion of CO into a metal–hydrogen bond. This does not mean, of course, that such a reaction, if it does take place at all, will be slow. Indeed, if such a reaction is fast, and is followed by other reactions that make up for the loss in bond energy, then it certainly could be the initial step in CO hydrogenation. We shall discuss this point in greater detail in Section 4.4. Finally, it should be noted that insertion reactions, as shown in Fig. 2.7, are *cis* in character.

2.3.3 β-Hydride Elimination

We have already seen in Section 2.2.2 that metal–alkyl compounds are prone to undergo β-hydride elimination or, in short, β-elimination reactions (see Fig. 2.5). In fact, hydride abstraction can occur from carbon atoms in other positions also, but elimination from the β-carbon is more common. As seen earlier, insertion of an alkene into a metal–hydrogen bond and a β-elimination reaction have a reversible relationship. This is obvious in Reaction 2.8. For certain metal complexes it has been possible to study this reversible equilibrium by NMR spectroscopy. A hydrido–ethylene complex of rhodium, as shown in Fig. 2.8, is an example. In metal-catalyzed alkene polymerization, termination of the polymer chain growth often follows the β-hydride elimination pathway. This also is schematically shown in Fig. 2.8.

2.3.4 Nucleophilic Attack on a Coordinated Ligand

Upon coordination to a metal center the electronic environment of the ligand obviously undergoes a change. Depending on the extent and nature of this change, the ligand may become susceptible to electrophilic or nucleophilic

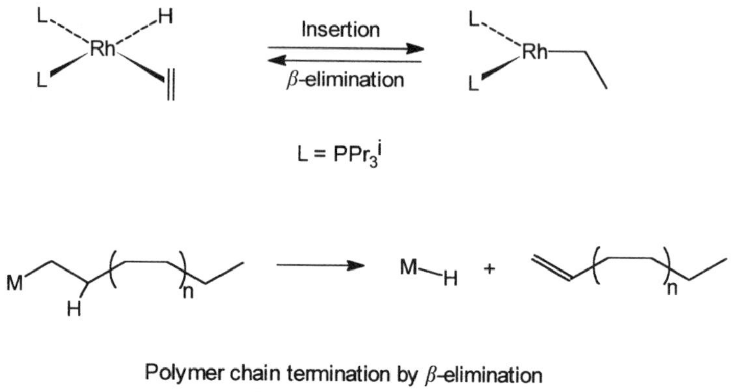

$$L = PPr_3{}^i$$

Polymer chain termination by β-elimination

Figure 2.8   Top: The relationship between insertion of an alkene into a metal–hydrogen bond and the reverse β-elimination reaction for a rhodium complex. Bottom: β-elimination leading to the formation of a metal hydride and release of a polymer molecule with an alkene end group.

attack. It is the enhanced electrophilicity or tendency to undergo nucleophilic attack that is often encountered in homogeneous catalytic processes. A few examples are shown by Reactions 2.12–2.14.

Nucleophilic attack by water on coordinated ethylene, as shown by Reaction 2.12, is the key step in the manufacture of acetaldehyde by the Wacker process (see Chapter 8). In Reaction 2.13 the high oxidation state of titanium makes the coordinated oxygen atom sufficiently electrophilic for it to be attacked by an alkene. As we will see in Chapter 8, this reaction is the basis for the homogeneous catalytic epoxidation of alkenes, using organic hydroperoxides as the oxygen atom donors.

(2.12)

(2.13)

(2.14)

The last reaction (2.14) has relevance as a model in the base-promoted water gas shift reaction, and is similar to Reaction 2.12. Instead of palladium-coordinated ethylene it is iron-coordinated carbon monoxide that undergoes attack by HO^-. The extent to which the reactivity of the ligand may be affected on coordination is often reflected in the rate constants. The ratio of the rate constants of nucleophilic attack by hydroxide on coordinated and free CO may be as high as 10^9!

2.4 ENERGY CONSIDERATIONS—THERMODYNAMICS AND KINETICS

For a chemical reaction to be experimentally observed, the thermodynamic and the kinetic changes should not be too unfavorable. The thermodynamic change is measured in terms of the change in Gibbs free energy, or simply free energy (ΔG), while the kinetic requirement is measured by free energy of activation ($\Delta G^{\#}$).

A thermodynamically favorable reaction ($\Delta G < 0$) may not take place because the free energy of activation may be too high. On the other hand, a thermodynamically unfavorable reaction ($\Delta G > 0$) may occur if the free energy of activation is low. There are many examples: The conversion of diamond to graphite is thermodynamically favorable but happens only at a vanishingly small rate at room temperature and pressure. A mixture of nitrogen and hydrogen does not automatically form ammonia; a considerable amount of energy has to be provided to overcome the activation energy barrier.

The relationship between ΔG and $\Delta G^{\#}$ is normally presented in a diagram, where free energies of the reactants, products, transition state, and intermediates are plotted against the extent of reaction, or more precisely the reaction coordinate. This is shown in Fig. 2.9. Even a simple homogeneous catalytic reaction such as alkene hydrogenation involves many intermediates and transition states. The free energy diagram thus resembles (c) rather than (a) or (b).

Finally, the relationship between equilibrium constant and free energy change in the standard state on the one hand, and rate constant and energy of activation on the other, are given by Eqs. 2.15 and 2.16, respectively. For calculating the $\Delta G^{\#}$ of a given reaction, $\Delta H^{\#}$ and $\Delta S^{\#}$ of the same reaction are calculated first. This is done by plotting $\ln(k/T)$ against $1/T$, where the slope and the intercept give the measures of $\Delta H^{\#}$ and $\Delta S^{\#}$, respectively. This type of a diagram is called an Eyring plot. A plot of $\ln k$ against $1/T$ is, of course, the Arrhenius plot and is used for measuring activation energy (ΔE)

$$-RT \ln K = \Delta G^0 \qquad (2.15)$$

$$k = Ae^{-\Delta E/RT} \qquad (2.16)$$

2.5 CATALYTIC CYCLE AND INTERMEDIATES

Consider a hypothetical metal complex ML_{n+1} (M = metal, L = ligand, $n + 1$ = number of ligands) that acts as a catalyst for the hydrogenation of an alkene. Also consider the following sequence of reactions:

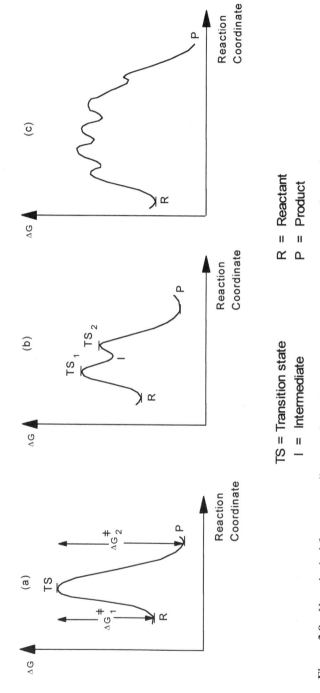

Figure 2.9 Hypothetical free energy diagrams. In (a) there is no intermediate (I). In (b) there is an intermediate (I) and $\Delta G = \Delta G_2^{\#} - \Delta G_1^{\#}$. The two transition states TS_1 and TS_2 would determine the kinetic behavior of the overall reaction. (c) Schematic depiction of a free energy diagram for an apparently simple catalytic reaction where a number of transition states and intermediates are involved.

$$ML_{n+1} \rightleftharpoons ML_n + L$$

$$ML_n + H_2 \rightleftharpoons H_2ML_n$$

$$H_2ML_n + \text{alkene} \rightleftharpoons H_2ML_n(\text{alkene})$$

$$H_2ML_n(\text{alkene}) \rightleftharpoons HML_n(\text{alkyl})$$

$$HML_n(\text{alkyl}) \rightarrow ML_n + \text{alkane}$$

If all these reactions excepting the first are added, we get the net stoichiometric reaction alkene $+H_2 \rightarrow$ alkane. The metal complex ML_n reacts with dihydrogen in the first step, undergoes a series of reactions in the following steps, and is regenerated in the final step. As shown in Fig. 2.10, all these reactions are conveniently presented as a *catalytic cycle*.

The following points are worth noting about the sequence of reaction presented in this cyclical manner. The stable metal complex added to the reaction

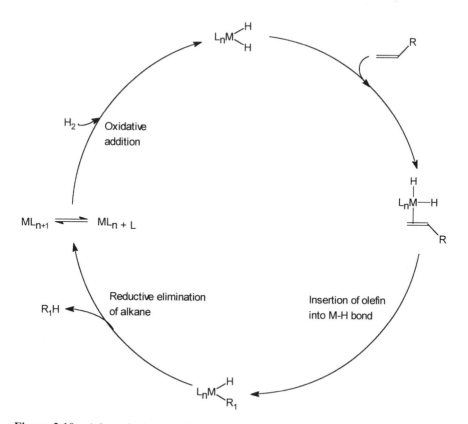

Figure 2.10 A hypothetical catalytic cycle with the precatalyst ML_{n+1} and four catalytic intermediates.

at the beginning, ML_{n+1}, is called the *precatalyst* or the catalyst precursor. All four metal complexes, ML_n, H_2ML_n, H_2ML_n(alkene), and HML_n(alkyl), that are present within the cycle are called *catalytic intermediates*. One complete catalytic cycle is derived from one molecule of the precatalyst and produces one molecule of the product. Turnover frequency in terms of a catalytic cycle is therefore the number of times the cycle is completed in unit time.

Information about the catalytic cycle and catalytic intermediates is obtained by four methods: kinetic studies, spectroscopic investigations, studies on model compounds, and theoretical calculations. Kinetic studies and the macroscopic rate law provide information about the transition state of the rate-determining step. Apart from the rate law, kinetic studies often include effects of isotope substitution and variation of the ligand structure on the rate constants.

Spectroscopic studies may be carried out under the actual catalytic conditions. These are referred to as in situ spectroscopic investigations. However, if the catalytic conditions are too drastic, it may not be possible to record spectra under such conditions. In such cases spectroscopic monitoring is done under less severe conditions.

Both kinetic studies and spectroscopic investigations have certain inherent limitations. Kinetic studies are informative about the slowest step, and at best can provide only indirect information about the fast steps. Spectroscopic detection of a complex, catalytically active or not, requires a minimum level of concentration. It is possible that the catalytically active intermediates never attain such concentrations and therefore are not observed. Conversely, the species that are seen by spectroscopy may not necessarily be involved in the catalytic cycle!

However, in most cases a combination of kinetic and spectroscopic methods can resolve such uncertainties to a large extent. The third method is based on the study of model compounds. Model compounds are fully characterized metal complexes that are assumed to approximate the actual catalytic intermediates. Studies on the reactions of such compounds can yield valuable information about the real intermediates and the catalytic cycle. With the advent of computational speed and methods, quantum-mechanical and other theoretical calculations are also increasingly used to check whether theoretical predictions match with experimental data.

We now discuss a few examples where these methods have yielded results that are particularly instructive. It must, however, be remembered that in most cases a combination of more than one method is necessary for an understanding of the observed catalysis at a molecular level.

2.5.1   Kinetic Studies

Dependence of reaction rates on the concentrations of the reactants (and the products in certain cases) can be very useful in understanding the mechanism of catalysis. Some of the ubiquitous mechanistic steps reveal themselves in empirically derived rate expressions. In other words, such rate expressions, once established by experiments, are characteristic signs of such steps.

One of the mechanistic steps most often encountered and inferred from kinetic data is ligand dissociation, which leads to the generation of a catalytically active intermediate. If ligand is added to such a catalytic system, the rate of the reaction decreases. Examples of this in homogeneous catalytic reactions are many: CO dissociation in cobalt-catalyzed hydroformylation, phosphine dissociation in RhCl(PPh$_3$)$_3$-catalyzed hydrogenation, Cl$^-$ dissociation in the Wacker process, etc. The actual rate expressions of most of these processes are described in subsequent chapters.

Another type of kinetic behavior that is very common for enzyme-catalyzed reactions (Michaelis–Menten kinetics) has also been observed in a number of homogeneous catalytic systems. The rate expression in such cases is given by

$$\text{rate} = kK[\text{substrate}][\text{catalyst}]/(1 + K[\text{substrate}]) \qquad (2.17)$$

where k is the rate constant, K is an equilibrium constant, and the square brackets signify concentrations. Kinetic behavior of this type is often called *saturation kinetics*. The physical significance of saturation kinetics is that a complex is formed between the substrate and the catalyst by a rapid equilibrium reaction. The equilibrium constant of this reaction is K, and it is then followed by the rate-determining step with rate constant k.

From the rate expression it is easy to see that increasing the substrate concentration will lead to an increase in rate initially, followed by a more or less constant rate at high substrate concentrations. The reason for this is that at high substrate concentrations $K[\text{substrate}] \cong 1 + K[\text{substrate}]$. Note that at constant catalyst concentration, a plot of (1/rate) against (1/[substrate]) will give a straight line. Saturation kinetics is observed in a number of homogeneous catalytic reactions such as hydrogenation, asymmetric hydrogenation, some epoxidation reactions, etc.

It must be remembered that with a change in reaction conditions a change in mechanism may also occur. What happens to be the rate-determining step under one set of reaction conditions need not necessarily be the rate-determining step under different conditions. A very good example of this is the Eastman Kodak process for methyl acetate carbonylation. Here there are two potential rate-determining steps. Which one of the two actually becomes slower obviously depends on the concentrations of the different reactants. This is discussed in detail in Section 4.6. Finally, as will be seen in subsequent chapters, there are many examples where isotope labeling and its effect on the rate or stereochemistry provide crucial mechanistic insights.

2.5.2 Spectroscopic Studies

Both infrared and multinuclear NMR spectroscopies have been used to identify homogeneous catalytic intermediates. These spectroscopic methods, if they are to be used for studying reactions under drastic conditions (i.e., high pressure and temperature), require careful choice of construction material and design of the spectroscopic cell.

A porphyrin complex of rhodium catalyzes the reactions between diazo-methane derivatives and alkenes to give cyclopropane rings. This is shown by:

$$R' CHN_2 \qquad N_2 \qquad (2.18)$$

A study of this reaction is an appropriate example of the usefulness of spectroscopic method in mechanistic studies. ^{1}H NMR studies on the reaction between the catalyst and the diazo compound show complexes 2.1 and 2.2 of Fig. 2.11 to be present in the reaction mixture. Complex 2.1 is stable at $-40°C$, and on warming to $0°C$ it undergoes conversion to 2.2. This reaction i.e., conversion of 2.1 to 2.2 is one of the several pieces of evidence for the intermediacy of the carbene complex 2.3. In other words, in situ NMR data, in conjunction with other evidences, indicate the involvement of 2.3 as a catalytic intermediate.

2.5.3   Model Compounds and Theoretical Calculations

Many compounds have been synthesized, characterized, and studied as models for proposed intermediates in various homogeneous catalytic reactions. Here we discuss two examples. Complex 2.4 is proposed as a model that shows the mode of interaction between an organic hydroperoxide and high-valent metal ions such as Ti^{4+}, V^{5+}, and Mo^{6+}. This type of interaction is considered to be necessary for the oxygen atom transfer from the hydroperoxide to an alkene to give an epoxide (see Chapter 8).

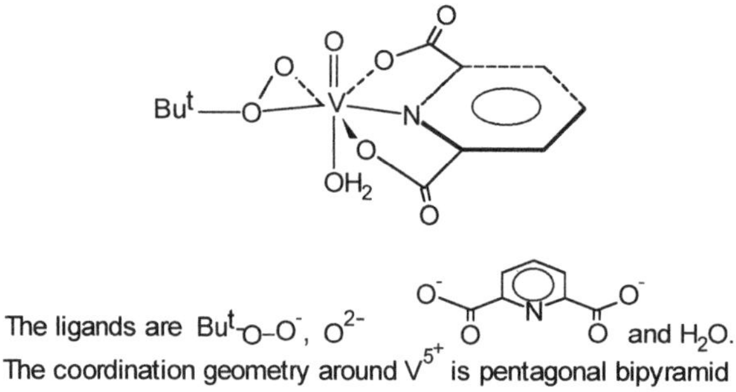

The ligands are But_{O-O}^-, O^{2-} and H_2O.
The coordination geometry around V^{5+} is pentagonal bipyramid

(2.4)

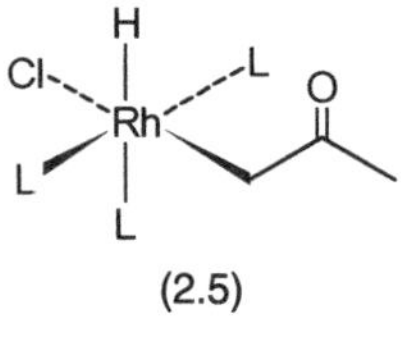

(2.5)

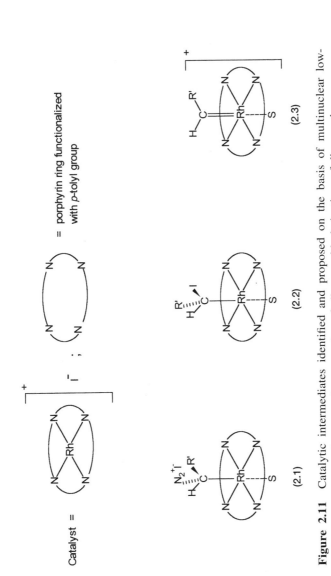

Figure 2.11 Catalytic intermediates identified and proposed on the basis of multinuclear low-temperature NMR studies on cyclopropanation of alkenes with derivatives of diazomethane.

The second example, 2.5, is a model for a hydrido–alkyl complex. This was studied structurally and kinetically to gain insight into the reductive elimination process that leads to the formation of an alkane. Complex 2.5 is synthesized by reacting $RhCl(PPh_3)_3$ with propylene oxide. This complex eliminates acetone to go back to $RhCl(PPh_3)_3$. Acetone, however, does not oxidatively add to $RhCl(PPh_3)_3$ to give 2.5, which is why this complex had to be made by the clever use of propylene oxide. The data obtained from kinetic studies on the acetone elimination reaction have been used to construct the free energy diagram shown in Fig. 2.12.

This diagram illustrates many important points. First of all, it shows the mechanistic complexity that may be anticipated for even an apparently simple reaction, reductive elimination of acetone from 2.5. Second, it shows that $RhCl(PPh_3)_3$ is thermodynamically more stable than complex 2.5 by about 40 kJ/mol. However, complex 2.5 does not undergo spontaneous conversion to $RhCl(PPh_3)_3$ because it has sufficient kinetic stability (>92 kJ/mol). Third, the high free energy of activation is associated with a ligand dissociation step that precedes the reductive elimination step. The five-coordinated intermediate, once

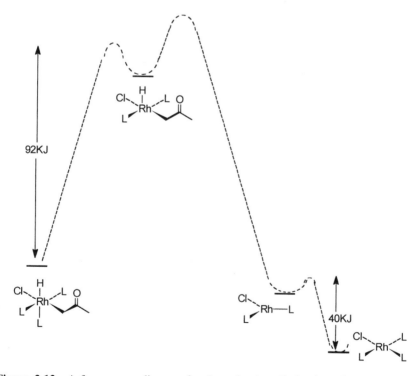

Figure 2.12 A free energy diagram for the reductive elimination of acetone from a model complex. Note the high activation energies required for the forward and the backward reactions.

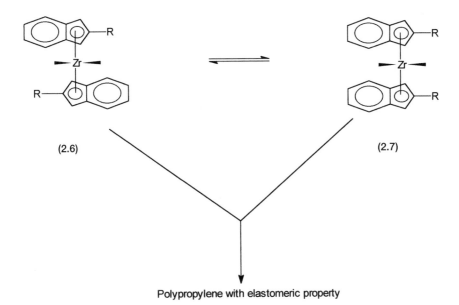

Figure 2.13   The energy difference between the stereomeric forms 2.6 and 2.7 of the metallocene catalyst determines the elastomeric property of the final polypropylene. The energy difference depends on the R group and can be estimated by theoretical calculations.

produced, requires little free energy of activation to eliminate acetone reductively, and form the three-coordinate $RhClL_2$. The latter adds a phosphine to form $RhCl(PPh_3)_3$.

Finally, we discuss one example that shows the importance of theoretical calculations in the design and understanding of a homogeneous catalytic process. Metallocene catalysts (Chapter 6) are increasingly used for making polymers with tailor-made properties. As shown in Fig. 2.13, one of these metallocene catalysts can exist in two stereochemical states, 2.6 and 2.7. The elastomeric property of the polypropylene, obtained by using this catalyst, depends on the relative amounts of 2.6 and 2.7. The energy required for the interconversion between the two states depends on the R group and can be theoretically calculated. On the basis of such calculations, for a variety of metallocene catalysts with different R groups, the elastomeric property of the final polypropylene has been correctly predicted.

PROBLEMS

1. What are the oxidation states and the valence electron counts of the metal ions in the complexes shown in Figs. 2.4, 2.6, and 2.11?

Ans. Fig. 2.4: 12, 10, 18, 16; Fig. 2.6: 16, 18 (2.2), 16, 18 (2.3), 16, 18 (2.4); Fig. 2.11: catalyst 14, assuming porphyrin to be a six-electron donor (covalent model), (2.1) to (2.3), all 18.

2. (a) In Fig. 2.2, which phosphine has a larger cone angle, and between R and R′ which one is more bulky? (b) In Fig. 2.6, Reaction 2.3, the six-coordinated rhodium complex may be expected to undergo what further reactions? (c) Draw schematic diagrams to show how bonding occurs in Ti^{4+}–alkene and Pt^{2+}–alkene. (d) It is difficult to isolate $W(C_2H_5)_6$ but not $W(CH_3)_6$. Why?

Ans. (a) PR_3' and R′. (b) Alkene insertion into Rh–H bond followed by reductive elimination of alkane. (c) No back-donation from $Ti^{4+}(d^0)$ but back-donation from $Pt^{2+}(d^8)$. (d) Facile β-elimination from C_2H_5.

3. The rate of displacement of X^- from $CH_3X(X = Cl, Br, I)$ by CN^- is in the order $I^- > Br^- > Cl^-$. What would be the expected order for reactions between RX and $[Rh(CO)_2I_2]^-$?

Ans. $I^- > Br^- > Cl^-$ assuming an S_N2 mechanism (two electrons from rhodium go to RI; i.e., rhodium acts as a nucleophile) for oxidative addition.

4. Postulate an intermediate for Reaction 2.5 with justification.

Ans. CpIr(CO), a 16-electron intermediate.

5. In Section 2.5 only the last reaction is not reversible. Why?

Ans. Oxidative additon of alkane to metal center is energetically unfavorable.

6. A soluble catalyst, 5 mmol, gives product, 3000 mmol, in 10 min. What is the turnover frequency and number of repeats of the catalytic cycle per second?

Ans. Both $1\ s^{-1}$.

7. In Fig. 2.10 ignore the ligand dissociation step and assume the oxidative addition of dihydrogen to be *a rapid reversible equilibrium* followed by the rate-determining step of alkene insertion. What is the expected rate expression?

Ans. Rate = kK[alkene][ML_n][H_2]/(1 + K[H_2]) when [H_2] >> [ML_n]. See Section 2.5.1.

8. In a homogeneous catalytic reaction addition of external ligand changes the rate as: [ligand] = 0.01, 0.005, 0.0033, and 0.0025 mmol/l, rate = 5, 10, 15, 20 mmol/(l, min). What will be the rate with a very large excess (~1 mmol/l) of ligand?

Ans. Plot rate against 1/[ligand] and calculate the intercept.

9. In Reaction 2.12 what organic product may be expected if β-elimination follows the nucleophilic attack by water on coordinated ethylene? What

condition must be satisfied for this to be a viable mechanism for catalytic conversion of ethylene to the product?

Ans. Vinyl alcohol, i.e., acetaldehyde. $[Pd-H]^+$ must be converted back to $[Pd-C_2H_4]^{2+}$.

10. Based on the reactions and complexes discussed in this chapter, what would be the expected products in the following reactions: (a) CH_3COI + $[Rh(CO)_2I_2]^-$; (b) Cp_2Lu-CH_3 + $CH_3CH{=}CH_2$; (c) $RuCl_2(PPh_3)_3$ + $ArCHN_2$; (d) $RhH(CO)(PPh_3)_2$ + $CH_3CH{=}CH_2$; (e) product of (d) plus CO; (f) product of (e) plus H_2.

Ans. (a) (4.4); (b) (6.16); (c) (7.44); (d) (5.4); (e) (5.6); (f) (5.7).

11. The rare earth metal complex Cp_2LuH catalyzes the hydrogenation of alkenes and on reaction with methane gives $Cp_2Lu(CH_3)$ and H_2. Suggest probable mechanisms.

Ans. *Not* by oxidative addition of dihydrogen. Polarizations of bonds, including the C–H of methane by direct interaction between the substrate and metal ion.

12. By referring to Fig. 2.12, what quantitative predictions could be made about (a) oxidative addition of acetone to $RhCl(PPh_3)_3$ and (b) free energy of the activation of the reaction between propylene oxide and $RhCl(PPh_3)_3$ to give (2.5).

Ans. (a) $\Delta G^{\#} > 132$ kJ/mol, (b) $\Delta G^{\#}$ considerably less than 132 kJ/mol.

13. By assuming insertion of alkene into metal–hydrogen bond to be the rate-determining step, draw a hypothetical free energy diagram for the catalytic cycle of Fig. 2.10. How many catalytic intermediates and transition states are there?

Ans. See Fig. 2.9(c). A similar diagram with four intermediates and four transition states.

BIBLIOGRAPHY

Sections 2.1 to 2.3.4

Books

Inorganic Chemistry, D. Shriver, P. Atkins, and C. H. Langford, W. H. Freeman, New York, 1994.

Advanced Inorganic Chemistry, F. A. Cotton and G. Wilkinson, Wiley, New York, 1988.

Principles and Applications of Organotransition Metal Chemistry, J. P. Collman and L. S. Hegedus, University Science Books, Mill Valley, California, 1987.

The Organometallic Chemistry of the Transition Metals, R. H. Crabtree, Wiley, New York, 1994.

Molecular Chemistry of the Transition Elements, F. Mathey and A. Sevin, Wiley, 1996.

Organometallic Chemistry, G. O. Spessard and G. L. Meissler, Prentice Hall, New Jersey, 1996.

Organometallics: A Concise Introduction, Ch. Elschenbroich and A. Salzer, VCH, Weinheim, 1989.

Articles

For oxidative addition of methane to cyclopentadiene iridium complexes, see: J. K. Hoyano and W. A. G. Graham, *J. Am. Chem. Soc.* **104**, 3723–25 (1982).

For nucleophilic attack on coordinated, CO see: P. C. Ford, *Acc. Chem. Res.* **14**, 31–37 (1981).

For reversible insertion and β-elimination, see D. C. Roe, *J. Am. Chem. Soc.* **105**, 7770–71 (1983).

Sections 2.4 and 2.5

Books

Physical Chemistry, P. W. Atkins, OUP, Oxford, 1995.

Kinetics and Mechanism of Reactions of Transition Metal Complexes, R. G. Wilkins, VCH, Weinheim, 1991.

Kinetics of Chemical Processes, Michel Boudart, Butterworth-Heinmann, London, 1991.

Articles

For the application of in situ IR in mechanistic studies, see R. Whyman, *Chemtech* **21**, 414–19 (1991).

For the application of in situ NMR in rhodium porphyrin-catalyzed cyclopropanation, see J. L. Maxwell et. al, *Science* **256**, 1544–47 (1992).

For model complexes (2.4) and (2.5), see H. Mimoun et al., *Nouv. J. Chim.* **7**, 467–75 (1983), and D. Milstein, *J. Am. Chem. Soc.* **104**, 5227–28 (1982), respectively.

For structural data on other model complexes, see N. L. Jones and J. A. Ibers, in *Homogeneous Catalysis with Metal Phosphine Complexes*, edited by L. H. Pignolet, Plenum Press, New York, 1983, pp 111–35.

For the application of theoretical calculations in metallocene-based polymerization, see J. T. Golab, *Chemtech* **28**(4), 19–29 (1998).

CHAPTER 3

CHEMICAL ENGINEERING FUNDAMENTALS

Commercial success of an industrial catalytic process depends on the feedstock cost, number of process steps, operability, reaction conditions, etc. A successful industrial process must strike the right balance between the selectivity and activity of the catalyst on the one hand, and its price and life on the other. Also, society demands that the manufacturing processes of the twenty-first century are inherently safe and environmentally friendly.

There are three main engineering factors that play an important part in translating a homogeneous catalyst-based laboratory process to a successful commercial one. These are reactor design, operating conditions, catalyst separation, and its reuse. In this chapter we give an outline of these principles. We also introduce aspects of current practices for maximizing safety and a favorable process economics. We begin our discussion by listing in Table 3.1 the functions of the various units of a typical process plant, the phases involved in such operations, and conditions under which these operations are usually carried out. Since heterogeneous catalytic processes play an overwhelmingly important role in chemical industry (see Sections 1.2 and 1.4), unit operations relating to heterogeneous catalytic reactions are also included.

3.1 REACTOR DESIGN

Homogeneous catalytic reactions are usually carried out in either a stirred tank or a tubular reactor. As can be seen from Table 3.1, tubular reactors may be of different types. Stirred tank reactors offer a very high degree of back mixing, while tubular reactors lead to plug flow conditions. The concentration of the reactants are uniform throughout the vessel in the case of stirred tank reactor,

TABLE 3.1 Principal Units of a Process Plant

Unit name	Function	Phases	Normal operating conditions
Reactor	Reaction	1. Gas–liquid–solid or gas–liquid	0–300 psi, 50–200°C
1. Stirred tank		2. In most cases solid with gas and/or liquid	
2. Tubular			
a. packed-bed			
b. Trickle-bed			
c. fluidized bed			
d. bubble column			
Crystallizer	Solid recovery from supersaturated solution	Liquid–solid	Atmospheric pressure, −10–100°C
Liquid–liquid extractor	Solute recovery	Liquid	Atmospheric pressure, −10–100°C
Nutshe-filter/centrifuge	Recovery of solids	Liquid–solid	Vacuum to atmospheric pressure, −10–100°C
Distillation column	Separation of liquids and purification	Liquid	Vacuum to 100 psi, −10–200°C
Absorption column/scrubber	Separation/recovery of gases	Gas–liquid	Atmospheric to 50 psi, 30–100°C
Heat exchanger	Cool or heat fluids	Gas or liquid	Atmospheric to 50 psi, −10–200°C
Condenser	Condense vapor to liquid	Gas–liquid	Atmospheric to 100 psi, −10–120°C
Evaporator/boiler	Vaporize liquid	Gas–liquid	Atmospheric to 100 psi, 30–200°C

while it keeps decreasing along the length of the tube in a tubular reactor. Note that for most heterogeneous catalytic reactions stirred tank or tubular reactors are the reactors of choice. In the following sections we discuss a few fundamental characteristics of these reactors.

3.1.1   Stirred Tank Reactors

The main issues to be considered in these reactors are good mixing, efficient gas dispersion in the liquid medium, and uniform heat transfer. The reactor can be operated in the batch or continuous mode. In the batch mode the reactants and the catalyst are charged, the reactor is heated to the desired temperature, and, after the completion of the reaction, the products are cooled and discharged. In the continuous mode of operation (see Fig. 3.1), the reactant and the catalyst mixture is continuously charged. The mixture of the product and

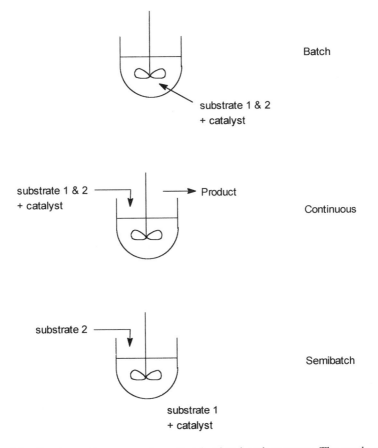

Figure 3.1   Batch, continuous, and semibatch stirred tank reactors. The mode of operation is schematically indicated.

the spent catalyst is also continuously discharged. A stirred tank reactor may also be operated in the semibatch (or semicontinuous) mode, where the liquid substrates and catalyst are charged at the beginning, while a gas or a second reactant is introduced continuously.

The manufacture of linear low-density polyethylene (LLDPE) by slurry polymerization in hexane (see Sections 6.2 and 6.8) is carried out by Hoechst, Mitsui, and a number of other chemical manufacturers in a series of continuous stirred tank reactors. The manufacture of butyraldehyde from CO, H_2, and propylene using a soluble rhodium phosphine complex (see Sections 5.2 and 5.5) is also carried out in a continuous stirred tank reactor.

Both batch and continuous stirred tank reactors are suitable for reactions that exhibit pseudo-zero-order kinetics with respect to the substrate concentration. In other words, under operating conditions the rate is more or less independent of the concentration of the substrate. However, for reactions where pseudo-first-order kinetics with respect to the concentrations of the substrates prevail, a batch tank reactor is preferred. Batch tank reactors are also ideally suited when there is a likelihood of the reactant slowly deactivating the catalyst or if there is a possibility of side product formation through a parallel reaction pathway.

A stirred tank reactor is a cylindrical container with a height-to-diameter ratio of 1:3, and it is often provided with baffles in order to avoid the formation of vortices during agitation. In general, four baffles with a width of 0.1 times the reactor diameter are arranged symmetrically with respect to the stirrer shaft. A sparger is provided at the bottom to introduce gas to the liquid.

Stirred tank reactors are provided with a jacket or immersion coil for heating or cooling the reaction medium. The temperature of the medium inside the tank is generally uniform. The rate of heat transfer depends on the heat transfer area, the difference in the temperature between the reaction medium and heating or cooling fluid and heat transfer coefficient.

Numerous types of stirrers are used in practice, and Fig. 3.2 shows the most commonly used ones. The stirrers for low-viscosity media are typically marine propeller and pitched-blade turbine stirrers, which cause axial fluid motion, and flat-blade turbines and impellers, which generate radial fluid motion. The former stirrers are suitable for uniformly suspending solids. The latter type are the preferred ones for carrying out exothermic reactions, like autooxidation reactions (see Section 8.3), where the heat generated during the process has to be removed effectively through the reactor walls.

In polymerization reactions viscosity builds up as the molecular weight increases. For such highly viscous media, helical stirrers that bring about axial motion or anchor stirrers are used. The latter also brings a scrapping motion, which helps in cleaning the reactor walls.

Power consumption by the stirrer depends on the dimensions of the stirrer, number of rotation of the stirrer per minute (rpm), viscosity, and density of the medium. Power requirements decrease when gas is bubbled through the liquid, as in hydrogenation and carbonylation reactions. Indeed in the Eastman Kodak

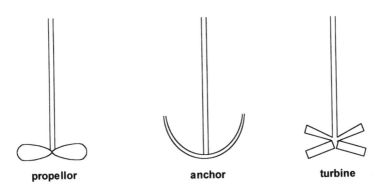

propellor　　　　**anchor**　　　　**turbine**

Figure 3.2　Three typical stirrers used in stirred tank reactors. Many other types are known.

acetic anhydride process (see Section 4.6) mixing is carried out by bubbling excess carbon monoxide and without any mechanical agitation. Gas dispersion also depends on the type of stirrer. Propeller stirrers are suitable for low to moderate gassing, while flat-blade turbines are ideal for high gassing.

Agitation or mixing brings the substrates nearer to each other so that they can react. There are two types of mixing, namely, macro- or bulk mixing and micromixing. Bulk mixing is dependent on vessel and agitator geometry, agitator speed, and solution viscosity. Also, the mixing time is inversely proportional to stirrer rpm. For example, in a vessel of 40 m^3 capacity having an agitator running at 25 rpm, the mixing time will be of the order of 30–60 sec. Micromixing, on the other hand, is a microscopic phenomenon. It arises due to molecular diffusion and is independent of mechanical agitation. The time scale for homogenization via molecular diffusion is of the order of 0.1–1 sec. In batch reactors, bulk mixing rather than micromixing effects is more common. If the bulk mixing time is greater than the time constant (inverse of the rate constant) of any side reaction, then desired product selectivity becomes poor.

3.1.2　Tubular Reactors

As shown in Table 3.1, tubular reactors could be of different types. A very large number of heterogeneous catalytic reactions are carried out in a tubular reactor of one type or another. In tubular reactors the reactant is fed in from one end of the tube with the help of a pump, and the product is removed from the other end of the tube. The other reactants, if there are any, may be introduced either at the reactor entrance in a cocurrent fashion, or from the reactor exit in a countercurrent fashion (Fig. 3.3).

The tube is generally packed with solid catalyst and/or inert packing to force good mixing and intimate contact between the reacting liquids and the catalyst. These reactors have simple construction and are always operated in the continuous mode. These reactors are the preferred ones when there is a likelihood

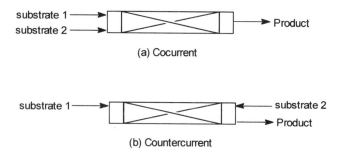

Figure 3.3 Tubular reactor showing cocurrent and countercurrent modes of operation. (The inner rectangle with diagonals represents a bed of catalyst.)

of catalyst deactivation due to a product or if side product formation follows a series reaction pathway (i.e., if the product reacts further to form unwanted side product).

Exxon and Phillips manufacture polypropylene in tubular reactors where the monomer is in the liquid form (see Section 6.8.2). One of the manufacturing processes for polyethylene involves the use of a "loop reactor" that has a recycle configuration. Here, under elevated pressure and temperature, a mixture of the catalyst, comonomer, hydrogen, and a solvent are introduced from one end of the reactor. The product and the unreacted starting materials are collected at the other end, and recycled back into the reactor.

In tubular reactors, heat is removed either by providing cooling tubes running parallel inside the reactor, or through external heat exchangers. Radial temperature gradients are normally observed in tubular reactors. In the case of exothermic reactions the temperature is maximum at the center of the tube and minimum at the tube wall. Similarly, in the case of endothermic reactions the temperature is minimum at the center of the tube and maximum at the tube wall. For highly exothermic reactions, packed bed reactors are usually avoided.

In packed bed reactors the solid catalyst is held stationary by plates at the top and bottom of the bed. In contrast, in fluidized bed reactors, the catalyst bed is relatively loosely packed, and there is no plate at the top. Rapid fluid flow from the bottom raises the bed and ensures good mixing, leading to insignificant temperature or concentration gradients. However, due to high fluid velocity some catalyst carryover is common.

3.1.3 Membrane Reactors

Another type of reactor that may have considerable future potential for use in homogeneous catalytic reactions is called the *membrane reactor*. These reactors have been successfully used for the commercialization of manufacturing processes based on enzyme catalysis. In fact, 75% of the global production of L-methionine is performed in an enzyme reactor. A membrane is basically an insoluble organic polymeric film that can have variable thickness. The catalyst or the enzyme may be bound inside the membrane or on its surface. In the latter case the membrane prevents the catalyst or the enzyme to pass through, but allows the migration of the products of the reaction. One of the main advantages of a membrane reactor is that it is suitable for carrying out both reaction and separation. The separation is based on molecular dimensions. The membrane material acts either as a catalyst or as a physical barrier. In the former case the membrane material itself may have catalyst activity, or the catalyst may be immobilized within the membrane.

3.1.4   Construction Materials

An important concern in industrial processes is corrosion. Transition metal complexes under certain conditions can facilitate corrosion of the reaction vessels. The consequences are not only fouling of the surfaces but also loss of expensive catalyst. So the reactors are generally made of materials that are resistant to corrosion. Two such materials are stainless steel 316 (containing 16–18% Cr, 10–14% Ni, 1–3% Mo, <0.1% C, and the rest iron) and Hastelloy C (containing 14–19% Mo, 4–8% Fe, 12–16% Cr, 3–6% W, and the rest Ni). The latter is ideal for chlorides and acids but is 3–4 times more expensive than the former. Wacker's process is operated under highly corrosive conditions (high concentrations of H^+ and Cl^-) (see Section 8.2); hence it requires expensive, titanium-lined reactors.

3.2   OPERATING CONDITIONS

For homogeneous catalytic reactions that involve at least one gas as one of the reactants, the reactor design requires the introduction of the reactant gases into a solution containing the catalyst. Since the reactions are generally exothermic, liberating roughly 120–210 kJ/mol of reactant, the well-mixed liquid medium usually helps rapid heat removal from the reactor. Operating conditions such as temperature, pressure, and solvent affect the selectivity of the products. At constant temperature, the solubility of gases in liquid medium increases with increasing operating pressure according to Henry's law. This is one of the reasons for operating gas–liquid homogeneous catalyzed reactions at high pressures, generally around 5–100 atm. Also, gas solubility decreases with increasing temperature.

Hydroformylation of linear olefins in a conventional cobalt oxo process (see Section 5.3) produces increasing linear-to-branched aldehyde ratios as the carbon monoxide ratio in the gas stream is increased up to 5 MPa (50 atm), but there is little further effect if the reaction mixture is saturated with carbon monoxide. An increasing partial pressure of hydrogen also increases this ratio up to a hydrogen pressure of 10 MPa. As the reaction temperature is increased, the linear-to-branched aldehyde ratios decreases. Solvents in conventional cobalt-catalyzed hydroformylation affect the isomer distribution. In propylene hydroformylation, dioxane and esters (for, e.g., butyl acetate) slightly increase the linear-to-branched aldehyde ratio compared to toluene, whereas the ratio decreases in the presence of acetone or diethyl ether. The reader may attempt to rationalize these effects of the operating conditions on the selectivity of the cobalt-based hydroformylation process on the basis of the molecular-level mechanism (see Section 5.3).

3.3 MASS TRANSFER IN MULTIPHASE REACTIONS

Gas–liquid multiphase catalytic reactions require the reacting gas to be efficiently transferred to the liquid phase. This is then followed by the diffusion of the reacting species to the catalyst. These mass transfer processes depend on bubble hydrodynamics, temperature, catalyst activity, physical properties of the liquid phase like density, viscosity, solubility of the gas in the liquid phase and interfacial tension.

In the case of a heterogeneous catalytic system consisting of gas, liquid, and solid catalyst, there are several factors that need to be considered for effective gas transfer (Fig. 3.4). These are:

1. Diffusion from bulk gas phase to gas–liquid interphase
2. Movement through the gas–liquid film
3. Diffusion through the bulk of the liquid region
4. Movement through liquid–catalyst film
5. Adsorption of the gas onto the catalyst surface

For homogeneous catalytic reactions the last two factors may be ignored. The rate of gas transfer into a liquid medium is given by $Q = k_l a(C_i^* - C_l)$, where k_l is the mass transfer coefficient, a, the interfacial area between the gas and liquid, C_i^* and C_l are equilibrium concentrations of the gas in the liquid phase and the concentration of gas in the liquid, respectively.

The gas-to-liquid transfer rate can be maximized by:

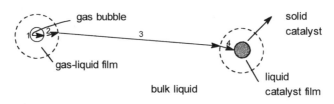

Figure 3.4 Different diffusion-related resistances in a gas–liquid–solid reaction. (Diffusion from bulk gas to the gas–liquid interface is represented by 1; movement through the gas–liquid film is 2, etc.)

1. Maximizing the interfacial area between the two phases
2. Increasing rate of transport of the gas molecules across the interface

A high interfacial area can be obtained by sparging the gas uniformly into the liquid phase in the form of small bubbles. Stirrer designs are available that aid in breaking the bubbles into small ones. High bubble breakage rates can be achieved by adding chemicals, that lower surface tension. Good gas–liquid contacts in tubular or column reactors can be achieved by packing the reactor with inert material of high surface area. Increasing the relative rate of flow of gas and liquid in the tubular reactor also aids in better mass transport. In stirred tank reactors, the rate of transport between the phases can be increased by selecting a suitable agitator, or by increasing the rpm of the stirrers.

A chemical reaction can proceed at a rate determined by intrinsic chemical reactivity only if the reactants are brought in contact at this rate. The rate of transport of the reactant molecules to the catalyst may be less than the rate of the reaction between them. In this situation, concentration gradients are set up, and the physical process of transport of the reactant to the catalyst, viz. by bulk diffusion, has an influence on the observed rate. In the homogeneous fluid phase the influence of bulk diffusion on the overall rate of reaction is often negligible. However, there are some fast reactions like proton transfer processes, which may be controlled by the rate of diffusion of the reactants. At higher temperatures, hydroformylation of linear olefins is controlled by carbon monoxide diffusion in the liquid phase. However, if stirring speeds are adequate to maintain carbon monoxide saturation in the liquid phase, the linear-to-branched ratios in the hydroformylation of propylene to 1-pentene and 2-pentene change only slightly over wide temperature ranges (see Section 5.2.2).

3.4 HEAT TRANSFER

Heating and cooling of reactants and products are generally carried out in tubular or coil heat exchangers. Two fluids at different temperatures flow in the tube and shell sides of the exchanger, and heat transfer takes place from the hot to the cold fluid. The rate of heat transfer (Q) is the product of the heat transfer area (A), the temperature difference between the two fluids (ΔT), and the overall heat transfer coefficient (U). In other words $Q = UA \ \Delta T$.

The overall heat transfer coefficient is a composite number. It depends on the individual heat transfer coefficients on each side of the tube and the thermal conductivity of the tube material. The individual heat transfer coefficient in turn depends on the fluid flow rate, physical properties of the fluid, and dirt factor. The temperature along the tube is not uniform. The hot and the cold fluids may flow in the same (cocurrent) or in opposite (countercurrent) directions. Generally the hot and cold fluids come in contact only once, and such an exchanger is called *single pass*. In a multipass exchanger, the design of the exchanger will permit the hot, and the cold fluids to change direction and come into close proximity with each other more than once.

3.5 CATALYST RECOVERY

In many homogeneous catalyst-based industrial processes efficient recovery of the metal is essential for the commercial viability of the technology (see Section 1.4). This is especially true for noble metal–based homogeneous catalytic reactions. Apart from economic reasons spent catalyst recovery is also essential to prevent downstream problems, such as poisoning of other catalysts, deposition on process equipment, waste disposal, etc. Several different techniques are being followed industrially for the recovery of the catalyst from the reaction medium after the end of the reaction:

1. Precipitation of the catalyst from the reaction medium, followed by filtration, as in the cobalt-based hydroformylation process (see Section 5.4). Here cobalt is removed from the reaction products in the form of one of its salts or as the sodium salt of the active carbonyl catalyst. The aqueous salts can be recycled directly, but sometimes they are first converted into an oil-soluble long-chain carboxylic acid salt, such as the corresponding naphthenate, oleate, or 2-ethylhexanoate.

2. Selective crystallization of the product. This leaves the catalyst and the residual substrate in the liquid phase. The liquid phase with the catalyst is normally recycled.

3. The use of a reverse osmosis membrane, which permits the chemicals to diffuse through, but prevents the catalyst from escaping. In hydrogenation and hydroformylation reactions, polystyrol-bound catalyst and phosphine are prevented from escaping with the liquid when polyamide membranes are used. As already mentioned, membrane technology for the recovery of homogeneous catalysts is still under development and has not yet reached commercial-scale applications. However, if successful, it would offer a clean and less energy-requiring separation technique.

4. Flash distillation of the product where very high vacuums are applied at moderate temperatures so the solvents and products vaporize, which are collected and condensed in a condenser, leaving the catalyst behind in the vessel. In the Monsanto acetic acid process, the catalyst rhodium iodide is left behind in the reboiler once the products are flashed off (see Section 4.9).

5. Liquid/liquid extraction of the catalyst, as in the DuPont adiponitrile process, where the nickel complex is extracted out of the product mixture after the reaction, with a solvent (see Section 7.7). In Shell's SHOP process the soluble nickel catalyst is also extracted from the reaction medium with a highly polar solvent, and reused (see Section 7.4.1).

6. Reaction in two-phase liquid–liquid systems. The Ruhrchemie process for the manufacture of butyraldehyde from propylene uses a water-soluble rhodium catalyst, while the product butyraldehyde forms an immiscible organic layer. Separation of the product from the catalyst is thus easily accomplished (see Section 5.2.5).

7. The chemical anchoring of complexes to a solid, such as silica-supported chromium catalyst, has been successfully used by Union Carbide for ethylene polymerization (see Section 6.2).

8. Confining the organometal in an organized structure like micelles, clay, or zeolite.

9. Reactions in supercritical environment.

The last two techniques are still being tested in the laboratory or pilot scale and have not so far been commercialized.

3.6   UNIT OPERATIONS

Product separation and catalyst recovery at the end of the homogeneous catalyzed reactions, as explained, are in most cases carried out by crystallization, filtration, distillation liquid–liquid extraction, or gas–liquid absorption. These unit operations can be performed in batch or continuous mode. The salient features of these operations are described in the following.

3.6.1   Crystallization and Filtration

Crystallization is the formation of solid particles from supersaturated liquid solution. Supersaturation is produced by the following ways.

1. Cooling of the mother liquor, such as in tank crystallizers
2. Evaporation of the solvent, such as in evaporator crystallizers
3. Evaporation combined with adiabatic cooling, such as in vacuum crystallizers

Solid particles are generally removed from the solution by filtration. The fluid is passed through a filtering medium, namely, a bed of fine particles, cloth, mesh, or sintered plate. The fluid flow is achieved either by forcing the mixture under pressure or by applying a vacuum on the opposite side.

3.6.2   Distillation

Separation of liquids that have different boiling points is carried out by distillation. Distillation may be carried out by two main methods. In the first method, known as *flash distillation*, the liquid mixture is converted into vapor by boiling it, usually under reduced pressure in a still, and condensing the resultant vapor in a receiver without allowing it to return to the still. This technique is used for separating components that boil at widely different temperatures or to recover catalyst as in the Monsanto acetic acid process.

In the second method a part of the condensate is returned (known as reflux) so that it comes in contact with the rising vapor on its way to the condenser. This technique is known as *rectification*. The efficiency of separation depends on the ratio of condensate returned to the still to the liquid product removed (which is known as the *reflux ratio*). The vapor and the reflux are made to come in intimate contact in a tall column located on top of the still. In the column the low boiler (liquid with a lower boiling point) spontaneously diffuses from the liquid reflux to the vapor, while the high boiler (the liquid with the higher boiling point) diffuses from vapor to liquid. So as the vapor rises in the column, it becomes enriched in low boiler and, as the liquid descends the column, its content of high boiler increases. A reboiler is provided at the bottom to vaporize the liquid and feed it up the column (Fig. 3.5).

An azeotrope is a liquid mixture that has the same composition both in the liquid and in the vapor phase. This means that the components cannot be separated by conventional distillation. In such cases an entrainer or a third component is added to make the compositions of the liquid and the gas phases different. In the case of liquid-phase butane autoxidation (Section 8.4), 2-butanone is separated as a pure component by adding entrainment agents such as ether to break the azeotrope the ketone forms with water.

A distillation column contains perforated plates or trays, or is stacked with inert packing material made of stainless steel, ceramic, or polymer. Distillation may be carried out either in a batch or in a continuous mode of operation. For the latter introduction of the feed into, and removal of the product from, the system is carried out continuously. A low-boiling solvent can be removed from

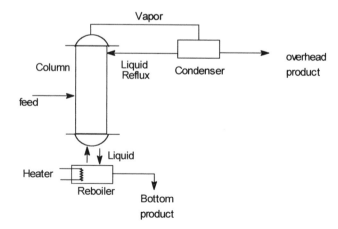

Figure 3.5 Continuous fractional distillation column.

the reaction mixture by steam stripping. Here live steam is introduced into the mixture, which carries the solvent with it to a condenser. This method is easy and is suitable for solvents that are immiscible with water.

3.6.3   Liquid–Liquid Extraction

In liquid–liquid extraction (Fig. 3.6) two miscible solutes are separated by a solvent, which preferentially dissolves one of them. Close-boiling mixtures that cannot withstand the temperature of vaporization, even under vacuum, may often be separated by this technique. Like other contact processes the solvent and the mixture of solutes must be brought into good contact to permit transfer of material and then separated. The extraction method utilizes differences in the solubility of the components in the solvent.

The extraction equipment consists of a mixer, where the solvent and the mixture of solutes are mixed in an agitated vessel, after which the two layers are made to separate in a settler tank. The equipment may be operated in batch or continuous mode. For batchwise extraction the mixer and the settler may be the same unit, whereas for continuous operation the mixer and the settler must be separate pieces of equipment. If several contact stages are required, a series

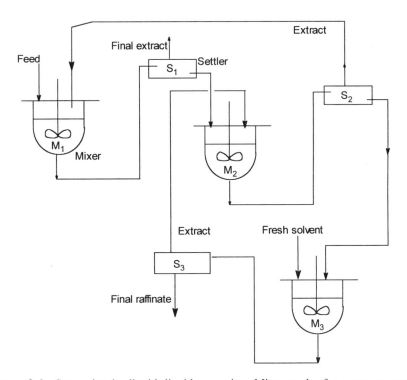

Figure 3.6   Separation by liquid–liquid extraction. Mixer settler for a countercurrent operation.

of mixers and settlers are placed in a continuous train and the two liquids are made to flow in cocurrent or in countercurrent fashion.

3.6.4   Gas–Liquid Absorption (or Scrubbing)

In this operation a soluble gas is absorbed from a gas mixture with a liquid. Generally the gas mixture and the liquid are brought in contact with each other in a packed column. Here the liquid flows from the top and the gas mixture rises from the bottom (countercurrent). The absorbed gas or vapor can be recovered later by desorption. The packing used in a gas–liquid absorption column is the same as the packing used in packed distillation columns and aids in good gas–liquid contact. This operation is also used to strip toxic vapors from exit gases. Such absorption columns are known as *scrubbers*.

3.7   SAFETY ASPECTS

The safety of the operating process, the health of the personnel handling the chemicals, and the impact of these chemicals on the environment have assumed major importance over the past two decades. Many modern techniques are available for identifying hazards and operability problems, and some of the common ones are systematic studies of hazardous operations (HAZOP), safety audits, and "what if" scenarios. Techniques like hazard analysis (HAZAN) help to quantify the frequency of occurrence of certain hazards based on a sequence of several individual events. HAZOP studies also help to plan waste disposal, transportation, storage, and handling of various chemicals used in the manufacturing process. Consequence analysis helps to determine the effect of certain events on the human habitation. Safety/hazard testing labs have cropped up all over the world that can be used for determining chemical toxicity, explosion hazards, and chemical incompatibilities.

Research is being directed towards developing processes that are inherently safe. Such processes must involve use of nontoxic chemicals, mild operating conditions, safe and a minimum number of solvents, insignificant side reactions, and low inventory. For example, if the inventory of methyl isocyanate, an intermediate in the Union Carbide plant in Bhopal, India, had been kept low, the accident would have been far less severe. Existing processes are being re-examined to decrease their severity. Also auditing of all chemicals from cradle to grave is now being followed in many large companies.

Homogeneous catalyst based carbonylation and hydroformylation reactions are generally carried out at high temperature and pressure. Metal carbonyls and carbon monoxide are toxic and require safe handling and disposal. The Wacker process for the manufacture of acetaldehyde from ethylene and oxygen needs appropriate safety measures (see Section 8.2), since the two gases can form an explosive mixture. This constraints the feed composition ratios and the conversion in the reactor. Although addition of nitrogen reduces the risk of explosion, the buildup of inerts leads to impurities in the product stream. Similarly, in auto-oxidation reactions peroxides are formed, and these may explode during their decomposition. So the rate of peroxide buildup and its decomposition are very carefully controlled within safety limits.

3.8 EFFLUENT AND WASTE DISPOSAL

Most homogeneous catalytic processes involve transition metals. The permissible limits for heavy metals in the effluent stream are less than 2 ppm, which means that the effluent treatment plant has to remove the transition metal from the effluent very effectively. The metal is normally precipitated in the form of a salt from the liquid stream, and then filtered off, or allowed to settle in settling tanks.

The recovery of heavy metals from solid waste poses more challenges. The Eastman Chemical Company process for the manufacture of acetic anhydride by the carbonylation of methyl acetate involves a proprietary process for the continuous recovery of rhodium and lithium from the process tar (see Section 4.6).

3.9 ECONOMICS

The cost of manufacturing a product includes the manufacturing cost, overhead, and general expenses. Manufacturing cost includes direct expenses and consists of the cost of raw material, containers, operator and labor costs, and utilities (like electricity, steam, water, fuel, etc.). This cost will depend on the production quantity. In contrast, overhead cost will be constant irrespective of the quantity of material that is being manufactured. Overhead cost may include expenses such as employee salaries, medical services, administration, insurance, depreciation, taxes, etc. General expenses consist of freight and delivery, sales, and R&D expenses.

A profitable product recovers these costs and leaves a margin (profit). As a rule of thumb, the selling price is usually between 1.5 to 3 times the raw material cost, namely, chemicals and catalyst. As already mentioned, in homogeneous catalytic processes the cost of the catalyst is an important component of the overall cost. The profit margin is usually low in the manufacture of bulk chemicals. Hence for the profitable manufacture of such chemicals the catalyst losses must be minimal.

The economics of any manufacturing process improves if the co-product or side product has a market. 90% of the world production of phenol is through the cumene hydroperoxide route because of the economic advantage of the co-product acetone. Oxirane technology for the production of propylene oxide from ethyl benzene leads to a co-product styrene and from isobutane leads to a co-product t-butyl alcohol.

PROBLEMS

1. A hot fluid enters a heat exchanger at a temperature of 150°C and is to be cooled to 60°C by a cold fluid entering at a temperature of 28°C and heated to 37°C. Will they be directed in cocurrent flow or countercurrent flow? Which flow can remove more heat?

Ans. Countercurrent.

2. Extraction followed by distillation is thought to be more economical than distillation alone in the case of separation of acetic acid from a dilute aqueous solution. State whether this statement is true or false.

Ans. True.

3. A 200-kg batch of solids containing 35% moisture is to be dried in a tray drier to 15% moisture content by passing hot air on its surface at a velocity of 2 m/s. If the constant rate of drying under these conditions is 0.85×10^{-3} kg/m^2 sec and critical moisture content is 15%, calculate the drying time. The material will be drying under constant rate from its initial value until it reaches a moisture content of 15%. The drying surface = 0.04 m^2/kg dry wt. Drying time = [(mass of dry solid)/(area × rate)] (initial moisture content − final moisture content).

Ans. 98 min

4. Heat transfer is taking place between two fluids in a shell and tube heat exchanger. (a) If one of the fluids is viscous, will it be directed in the tube or shell side? (b) If one of the fluids is corrosive, will it be directed in the tube or shell side?

Ans. (a) The viscous fluid will be directed in the shell side. The higher the viscosity, the higher will be the pressure drop (i.e., resistance to flow), and if it is directed in the tube side, the pressure drop will be still larger.
(b) The corrosive fluid will be directed in the tube side, and if any tube becomes corroded, it can be replaced.

5. An aqueous ethanol solution containing 8.6% ethanol is fed into a fractionating column at the rate of 1000 kg/h to obtain a distillate containing 95.4% alcohol, and at the bottom a solution of 0.1% alcohol. Calculate the mass flow rates of the distillate and bottoms.

Ans. Carry out a mass balance for ethanol:

$$\text{ethanol in} = \text{ethanol in distillate} + \text{ethanol in bottoms}$$

$$\text{bottom} = 910.8 \text{ kg/h, distillate} = 89.2 \text{ kg/h.}$$

6. Calculate the area of heat transfer of a shell and tube heat exchanger operated in a countercurrent fashion under the following conditions. Overall heat transfer $(U) = 4000$ kcal/h m^2 °C, heat duty $(Q) = 6,00,000$ kcal/h,

hot fluid inlet = 120°C; outlet = 90°C. Cold fluid inlet = 30°C, outlet = 43°C. $Q = UA\ \Delta T_{\ln}$, where $\Delta T_{\ln} = (\Delta T_2 - \Delta T_1)/\ln(\Delta T_2/\Delta T_1)$. For countercurrent ΔT_2 = temperature of hot fluid in temperature of cold fluid out. ΔT_1 = temperature of hot fluid out temperature of cold fluid in.

Ans. 2.2 m².

7. Repeat this problem for a cocurrent (parallel) flow shell and tube heat exchanger. Here ΔT_2 = temperature of hot fluid in − cold fluid in. ΔT_1 = temperature of hot fluid out − cold fluid out.

Ans. 2.6 m².

8. State the methods of avoiding vortex in a liquid under agitation. What are the disadvantages of vortex formation?

Ans. Vortex can be avoided (a) by providing baffles in the tank; (b) by placing the agitator off-center; and (c) by providing an extra set of flat-blade agitators at the surface of the liquid. The vortex leads to improper agitation and (b) ingress of gas from the space above the liquid due to the creation of a vacuum.

9. In a distillation column a vent is usually provided at the top of the condenser. What will happen if the rate of vaporization is increased above that for which the condenser is designed?

Ans. Uncondensed vapor will escape.

10. The driving force for heat transfer is the temperature difference. What is the driving force for mass transfer?

Ans. The concentration difference.

11. Ammonia is to be absorbed from a mixture of ammonia and air in an absorption tower (countercurrent) using water as the solvent. The conditions are as follows: Air flow rate (V) = 200 kg/h. Liquid-phase compositions: at the top of the packing (x_2) = 0.000013 kg NH_3/kg H_2O; at the bottom of the packing (x_1) 0.0006 kg NH_3/kg H_2O. Gas-phase composition: at the bottom of the packing (y_1) = 0.0084 kg NH_3/kg of gas; at the top of the packing (y_2) = 0.00044 kg NH_3/kg gas. Calculate the flow rate of water (L).

Ans. Carry out a mass balance for ammonia. Ammonia entering the column with the gas-ammonia leaving the column with the gas = ammonia leaving the column with the liquid–ammonia entering the column with the liquid. $V(y_1 - y_2) = L(x_1 - x_2)$. 1362.86 kg/h.

12. In a stirred vessel, power input (P) to ghe agitator is proportional to $N_i^2 D_i^3$ (in laninar regime) and $N_i^3 D_i^5$ in the turbulent region. During aeration power input (P_a) to the agitator is proportional to $[P^2(N_i D_i^3)^{0.44}/N_a^{0.56}]^{0.45}$ (N_i and D_i are the stirrer rpm and diameter, P, the power input in the absence

of aeration, $N_a = Q/N_iD_i^3$, Q the aeration flow rate). (a) How will motor power vary with aeration rate? (b) Compare motor power during aeration and without aeration for varying stirrer rpm and explain your results.

Ans. (a) Increasing aeration decreases power requirement for agitation; (b) rpm has a larger effect on P_a than P, since power is consumed in shearing the air bubbles.

BIBLIOGRAPHY

Books for all the sections

Unit Operations of Chemical Engineering, 5th ed., W. L. McCabe and J. C. Smith, McGraw-Hill, New York, 1993.

Process Heat Transfer, D. Q. Kern, McGraw-Hill, New York, 1950.

Unit Operations I and II, K. A. Gavhane, Nirali Prakashan, Pune, 1992.

Fluid Mixing Technology, Chemical Engineering, J. Y. Oldshue, McGraw-Hill, New York, 1983.

Article

M. Baerns and P. Claus, in *Applied Homogeneous Catalysis*, edited by B. Cornils and W. A. Herrmann, VCH, Weinheim, New York, 1996, vol **2**, 684–98.

CHAPTER 4

CARBONYLATION

4.1 INTRODUCTION

In this chapter we discuss the mechanistic and other details of a few industrial carbonylation processes. These are carbonylation of methanol to acetic acid, methyl acetate to acetic anhydride, propyne to methyl methacrylate, and benzyl chloride to phenyl acetic acid. Both Monsanto and BASF manufacture acetic acid by methanol carbonylation, Reaction 4.1. The BASF process is older than the Monsanto process. The catalysts and the reaction conditions for the two processes are also different and are compared in the next section. Carbonylation of methyl acetate to acetic anhydride, according to reaction 4.2, is a successful industrial process that has been developed by Eastman Kodak. The carbonylation of propyne (methyl acetylene) in methanol to give methyl methacrylate has recently been commercialized by Shell. The Montedison carbonylation process for the manufacture of phenyl acetic acid from benzyl chloride is noteworthy for the clever combination of phase-transfer and organometallic catalyses. Hoechst has recently reported a novel carbonylation process for the drug ibuprofen.

$$CH_3OH + CO \rightarrow CH_3CO_2H, \quad \Delta H = -136.6 \text{ kJ/mol} \quad (4.1)$$

$$CH_3CO_2CH_3 + CO \rightarrow (CH_3CO)_2O, \quad \Delta H = -94.8 \text{ kJ/mol} \quad (4.2)$$

In this chapter we also discuss two other catalytic reactions that involve CO as one of the reactants. These are the water-gas shift (see Section 1.2) and Fischer–Tropsch reactions. Although for these reactions homogeneous catalysts are not used industrially, for explaining the formation of by-products in the

Monsanto and BASF processes, they are relevant. Many of the fundamental concepts discussed in Chapter 2 are neatly illustrated by the methanol carbonylation reactions. We begin our discussion with these reactions.

4.2 MANUFACTURE OF ACETIC ACID

Some of the process parameters for methanol carbonylation processes are compared in Table 4.1.

The main byproduct forming reactions in the BASF and Monsanto processes are different. In the former it is the liquid phase Fischer–Tropsch-type reaction, that leads to the formation of products such as such as alkyl acetates, methane etc. In the Monsanto process it is the homogeneous water-gas shift reaction that produces CO_2 and H_2 as byproducts. Also note that the Monsanto process is superior in terms of selectivity, metal usage and operating conditions.

4.2.1 The Monsanto Process—The Catalytic Cycle

The basic catalytic cycles and the catalytic intermediates for the Monsanto process are shown in Fig. 4.1. A variety of rhodium salts may be added to the reaction mixture as precatalysts. In the presence of I^- and CO they are quickly converted to complex 4.1. The following points about the catalytic cycles deserve special attention.

First, the organometallic and organic catalytic steps are shown separately in (a) and (b), respectively. In Fig. 4.2 these two loops are combined. In this diagram there is also one extra catalytic intermediate (4.5). This is involved in

TABLE 4.1 Comparison of Process Parameters for Methanol Carbonylation Processes

	BASF	Monsanto
Metal concentration	10^{-1} mole/liter of cobalt	10^{-3} mole/liter of rhodium
Temperature	230°C	180–190°C
Pressure (atm)	500–700	30–40
Selectivity (%) based on		
1. Methanol	90	>99
2. CO	70	90
Byproducts	CH_4, glycol acetate, and other oxygenated hydrocarbons	CO_2, H_2
Effect of hydrogen	Amount of byproduct increases	No effect
Promoter CH_3I	Essential	Essential

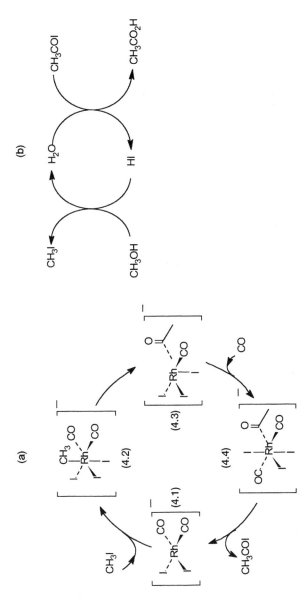

Figure 4.1 Monsanto process. (a) The organometallic catalytic cycle; CH_3I and CO react to give CH_3COI. (b) The organic catalytic cycle; water and HI act as catalysts to generate acetic acid and CH_3I from CH_3COI and methanol.

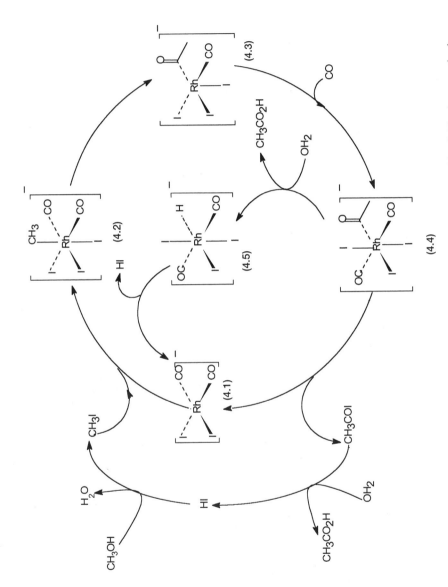

Figure 4.2 Monsanto process: The organic and organometallic cycles are combined. The inner cycle shows an additional pathway for product formation.

an additional product-forming pathway. Second, 4.1 to 4.2 is an oxidative addition reaction, 4.2 to 4.3 is an insertion reaction, and 4.4 to 4.1 is a reductive elimination reaction. Third, 4.1 and 4.3 are 16-electron complexes, while 4.2, 4.4, and 4.5 have electron counts of 18. Finally, the stoichiometry of the overall reaction is that of reaction 4.1; methanol and CO are the only two reactants that irreversibly enter the loops.

Acetyl iodide is the real product of the primary catalytic cycle. Water, though required for the hydrolysis of acetyl iodide, is generated in the reaction of methanol with HI. It is therefore not involved in the overall stoichiometry. To make the cycle operational, small amounts of CH_3I and water are added in the beginning.

4.2.2 Mechanistic Studies and Model Compounds

The rate of methanol carbonylation shows zero-order dependence on the concentration of methanol, carbon monoxide, and acetic acid. As shown, it is first order with respect to the concentrations of rhodium and methyl iodide:

$$rate = [Rh][CH_3I] \qquad (4.3)$$

The rate-determining step in the catalytic cycle is the oxidative addition of CH_3I to 4.1. Oxidative additions of alkyl halides are often known to follow S_N2 mechanisms. This appears to be also the case here. The net negative charge on 4.1 enhances its nucleophilicity and reactivity towards CH_3I.

In situ IR spectroscopy of the reaction mixture at a pressure and temperature close to the actual catalytic process shows the presence of only 4.1. As we will see, all the other complexes shown in the catalytic cycle (i.e., complexes 4.2–4.5) are also seen by spectroscopic methods under milder conditions. Under the operating conditions of catalysis they are not observed by IR spectroscopy, as their concentrations are much less compared to that of 4.1.

Regardless of the initial source of rhodium, that is, whether it is added just as a halide or as a phosphine complex, under the reaction conditions 4.1 is formed. With phosphine complexes, the phosphine is converted to a phosphonium (PR_4^+ or HPR_3^+) counter-cation. As chelating phosphines bind more strongly than the monodentate ones, the induction time for the formation of 4.1 with chelating phosphines is longer.

At ambient temperatures and in neat CH_3I as a solvent, IR bands and NMR signals for complexes 4.2–4.4 are seen. The stereochemistry of 4.2 as shown in the catalytic cycle is consistent with the spectroscopic data. The relative thermodynamic and kinetic stabilities of complexes 4.1–4.3 under these conditions have also been estimated. The data show that 4.2 is unstable with respect to conversion to both 4.3 and 4.1. In other words, 4.2 undergoes facile insertion and reductive elimination reactions.

The spectral characteristic of 4.3 in solution is indicative of the presence of a mixture of isomers in solution. From the reaction of CH_3I with 4.1, under noncatalytic conditions, a solid has been isolated. The structure of this compound is as shown in 4.6. It is clear that 4.6 is a dimer of 4.3. In other words, under noncatalytic conditions (absence of external CO), 4.1 reacts with CH_3I to give 4.3, which then dimerizes to produce the stable compound.

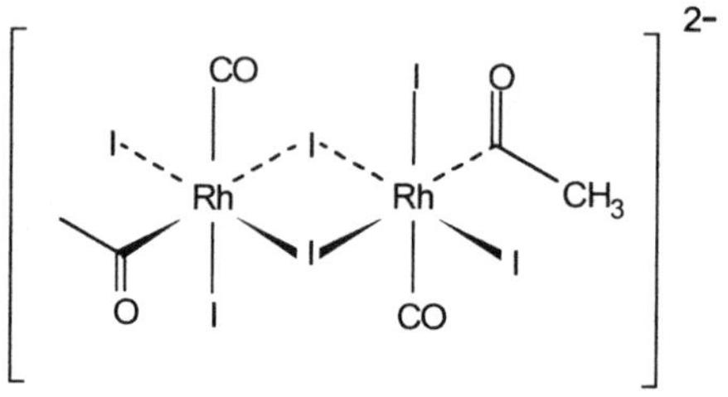

(4.6)

Reaction of 4.6 with CO has been investigated by spectroscopy (IR and NMR) and product analyses. The final products of this reaction are 4.1 and CH_3COI. In this conversion of 4.6 to 4.1, there is spectroscopic evidence to suggest the formation of an intermediate complex like 4.4 in solution. In fact, by carrying out low-temperature NMR studies on ^{13}C-labeled compound, the sequence as shown by 4.4, has been suggested.

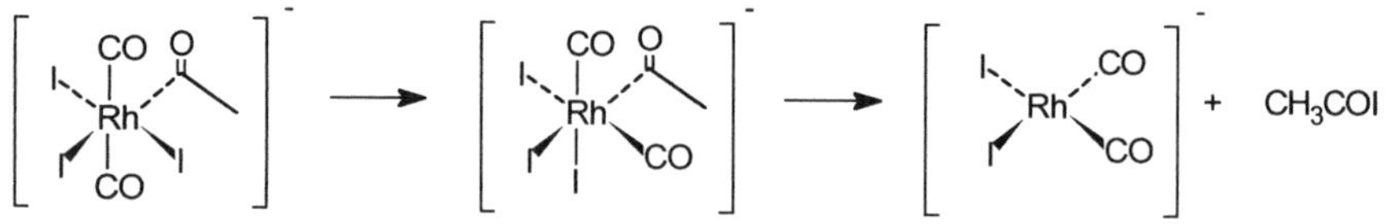

(4.4)

Finally, complex 4.5 can be made from the oxidative addition of HI onto 4.1. The suggestion that 4.4 reacts with water to form acetic acid and 4.5 is therefore a reasonable one. In summary, there is excellent evidence for all the catalytic intermediates shown in the proposed cycle.

4.2.3 The BASF Process—The Catalytic Cycle

The basic steps of the catalytic cycle with the cobalt catalyst are shown in Fig. 4.3. The tetracarbonyl cobalt anion 4.7 is formed from cobalt iodide, by reactions 4.5–4.7.

$$2CoI_2 + 2H_2O + 10CO \rightleftharpoons Co_2(CO)_8 + 4HI + 2CO_2 \tag{4.5}$$

$$Co_2(CO)_8 + H_2O + CO \rightleftharpoons 2HCo(CO)_4 + CO_2 \tag{4.6}$$

$$3Co_2(CO)_8 + 2n\ MeOH \rightleftharpoons 2[Co(MeOH)_n]^{2+} + 4Co(CO)_4^- + 8CO \tag{4.7}$$

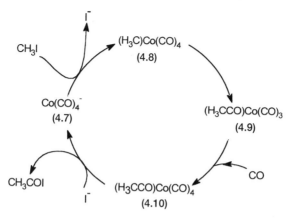

Figure 4.3 BASF process: The catalytic cycle for the conversion of CH_3I plus CO to CH_3COI. The organic cycle is not shown and is the same as in Figs. 4.1 and 4.2.

A nucleophilic attack by 4.7 on CH_3I produces 4.8 and I^-. Conversion of 4.8 to 4.9 is an example of a carbonyl insertion into a metal alkyl bond. Another CO group adds onto the 16-electron species 4.9 to give 4.10, which in turn reacts with I^- to eliminate acetyl iodide. Formation of acetic acid and recycling of water occur by reactions already discussed for the rhodium cycle. Apart from these basic reactions there are a few other reactions that lead to product and by-product formations. As shown in Fig. 4.4, both 4.9 and 4.10 react with water to give acetic acid. The hydrido cobalt carbonyl 4.11 produced in these reactions catalyzes Fischer–Tropsch-type reactions and the formation of by-products. Reactions 4.6 and 4.7 ensure that there is equilibrium between 4.7 and 4.11.

There are two main differences between cobalt- and rhodium-based catalytic cycles. In the cobalt-catalyzed cycle it is the nucleophilic attack by 4.7 on CH_3I rather than the oxidative addition of CH_3I on any unsaturated 16-electron species that initiates the catalytic cycle. Second, in the rhodium cycle reductive elimination generates acetyl iodide, whereas in the cobalt cycle it is the attack by I^- on 4.10 that produces acetyl iodide. Thus oxidative addition and reductive elimination steps are *not* involved in the cobalt cycle, but play crucial roles in the rhodium cycle.

4.2.4 BASF Process—Mechanistic Studies

The rate of cobalt-catalyzed carbonylation is strongly dependent on both the pressure of carbon monoxide and methanol concentration. Complex 4.7, unlike 4.1, is an 18-electron nucleophile. This makes the attack on CH_3I by 4.7 a comparatively slow reaction. High temperatures are required to achieve acceptable rates with the cobalt catalyst. This in turn necessitates high pressures of CO to stabilize 4.7 at high temperatures.

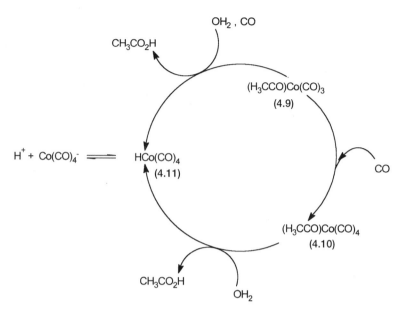

Figure 4.4 Pathways that lead to the generation of $HCo(CO)_4$ and an equilibrium between this species and $Co(CO)_4^-$.

Spectroscopic studies under actual operating conditions have not been reported. However, 4.7, 4.8, 4.10, and 4.11 are well-characterized complexes. Evidence for the intermediate 4.9 comes from kinetic studies. The structure of 4.11 has also been determined by electron diffraction. X-ray structural data are available for $Co(C_2F_4H)(CO)_3L$, a model of 4.8.

4.3   WATER-GAS SHIFT REACTION AND RHODIUM-CATALYZED CARBONYLATION

The relevance of the water-gas shift reaction in the petrochemical industry has already been discussed (see Section 1.1). The significance of the water-gas shift reaction in homogeneous systems is twofold. First, it plays a crucial role in stabilizing the rhodium catalyst in the Monsanto process. Second, studies carried out in homogeneous systems employing metals other than rhodium have provided useful mechanistic insights into the heterogeneous water-gas shift reaction. We first discuss the catalytic cycle with 4.1 as one of the catalytic intermediates, and then mechanistic results that are available from an iron-based catalytic system.

The proposed catalytic cycle for the water-gas shift reaction in the Monsanto process is shown in Fig. 4.5. This cycle operates at acidic pH and is responsible for CO_2 and H_2 production. It has a useful function in stabilizing the rhodium

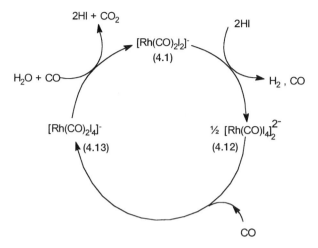

Figure 4.5   Proposed catalytic cycle for the water-gas shift reaction in the Monsanto process intermediates such as 4.5, 4.28, etc. may also be involved, but there is no direct evidence.

complex in the absence of CH_3I. In a situation where no CH_3I is available, the acetic acid–forming catalytic cycle (Fig. 4.1) ceases to exist. However, since the water-gas shift cycle continues to be operational, rhodium remains in solution and does not precipitate out.

At a molecular level, the intimate mechanism for the water-gas shift reaction with 4.1 as the precatalyst is not known in any detail. In the conversion of 4.1 to 4.12, the evolution of hydrogen is probably indicative of the intermediacy of a hydrido complex. Similarly, in the conversion of 4.13 to 4.1, where HI and CO_2 are eliminated, nucleophilic attack on coordinated CO may be involved. Although these mechanistic conjectures are plausible, there is no direct evidence to support them.

The water-gas shift reaction has also been studied under high pH with metal carbonyls as catalysts. The catalytic cycle with $Fe(CO)_5$ as the precatalyst is shown in Fig. 4.6. This reaction with low turnover is carried out at 130–180°C, under 10–40 bar of CO, with alkali metal hydroxide as a promoter.

In situ IR spectroscopy shows the presence of 4.14 and 4.16 in the catalytic mixture. The presence of 4.15 is inferred from the known chemistry of metal carbonyls. Both nucleophilic attacks on coordinated CO by HO^- and decarboxylation of the resultant complex as shown in 4.8 are well-precedented reactions.

$$M-CO + OH^- \longrightarrow \left[M-C{\overset{\displaystyle O}{\underset{\displaystyle O-H}{\diagdown}}} \right]^- \longrightarrow [M-H]^- + CO_2$$

$$(4.8)$$

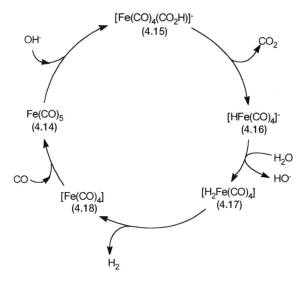

Figure 4.6 Proposed catalytic cycle for the water-gas shift reaction by $Fe(CO)_5$ at high pH.

4.4 FISCHER–TROPSCH REACTION AND COBALT-CATALYZED CARBONYLATION

The Fischer–Tropsch reaction, named after the original inventors, is essentially the conversion of synthesis gas $(CO + H_2)$ to a mixture of hydrocarbons, and to a lesser extent oxygenated hydrocarbons. The commercial Fischer–Tropsch reaction such as the one practiced at SASOL in South Africa uses potassium- and copper-promoted heterogeneous iron catalysts. A wide range of hydrocarbons, containing one to more than 100 carbon atoms, are produced. Paraffins and to a lesser extent alkenes are the main products. These are thought to arise through carbide intermediates that are converted to CH_2 groups. There is no homogeneous Fischer–Tropsch catalyst that gives paraffin or alkene in good yield.

The CO used for the carbonylation reaction always contains some hydrogen. The side products in the BASF carbonylation process arise due to Fischer–Tropsch reaction catalyzed by the cobalt catalyst. The high temperatures and pressures used in the BASF process are conditions under which the Fischer–Tropsch reaction with soluble cobalt catalyst can take place. In the Monsanto process the reaction conditions are much milder, and the side-product-forming Fischer–Tropsch reaction is avoided.

Based on the known reactions of cobalt carbonyls, the catalytic cycle shown in Fig. 4.7 has been suggested. The hydrido carbonyl 4.11 is converted to 4.19, a formyl species. Formyl complexes are considered to play a key role in all

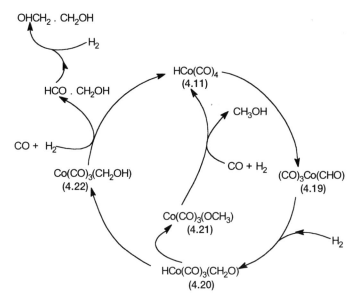

Figure 4.7 Proposed catalytic cycle for the Fischer–Tropsch reaction, giving oxygenated hydrocarbons.

homogeneous catalytic reactions that give mainly oxygenated hydrocarbons from synthesis gas. Intramolecular CO insertion into a metal–hydride bond is thermodynamically unfavorable under mild conditions (see Section 2.3.2). However, under high pressure and temperature this reaction may be kinetically favorable or may take place by an *intermolecular* mechanism. The formyl complex reacts with dihydrogen to give 4.20, a complex with coordinated formaldehyde. Two different reactions leading to the formation of 4.21 or 4.22 are possible. The hydroxymethyl complex 4.22 is responsible for the formation of ethylene glycol through the intermediate formation of glycol aldehyde. On the other hand, 4.21 produces methanol.

To this date most of the reactions, though reasonable from the point of view of established organometallic chemistry of cobalt and other metals, remain speculative. Formyl, formaldehyde, hydroxymethyl, and methoxy complexes of transition metals are known, and this to a large extent gives confidence about the correctness of the proposed catalytic cycle. For example, the production of ethylene glycol from 4.22 must proceed through a catalytic intermediate, where the carbon atom of the hydroxymethyl group forms a C–C bond with a carbonyl or a formyl group. Such complexes are indeed known and fully characterized for rhodium and iridium. As shown for 4.23, part of the stability of such complexes may be due to intramolecular hydrogen bonding.

$$\text{(M = Rh, Ir)} \quad \text{L = PPh}_3$$
$$(4.23)$$

In the 1970s Union Carbide had reported the use of rhodium with promoters such as amines, carboxylates, etc. for the synthesis of ethylene glycol from CO plus H_2. Manufacture of ethylene glycol by this route, however, was never commercialized. The mechanism of this reaction is not understood. Both mononuclear and polynuclear (cluster) rhodium carbonyls can be seen by NMR and IR spectroscopy under conditions approximating that of the catalytic reaction. The question as to whether the catalytic intermediates are mononuclear or cluster has not been answered with any certainty so far.

Some mechanistic information is available on ruthenium-based homogeneous Fischer–Tropsch reactions. By in situ IR spectroscopy, in the absence of any promoter, only $Ru(CO)_5$ is observed. An important difference between the cobalt and the rhodium system on the one hand and ruthenium on the other is that in the latter case no ethylene glycol or higher alcohols are obtained. In other words, in the catalytic cycle the hydroxymethyl route is avoided.

With the ruthenium-based catalyst, in the presence of a promoter such as I^- both activity and selectivity towards oxygenated two-carbon-containing products increase. Under these conditions in situ IR spectroscopy shows the presence of the polynuclear complex $[HRu_3(CO)_{11}]^-$ and the mononuclear $[Ru(CO)_3I_3]^-$. These complexes themselves are not catalytically active intermediates; rather they are the precursors of such intermediates. From independent experiments it appears that $[HRu(CO)_4]^-$ and $[Ru(CO)_4I_2]$ are two of the actual active intermediates that are formed from the cluster and the mononuclear complexes, respectively. These complexes are not detected by spectroscopy, presumably because of insufficient concentration. There is also evidence to suggest that $[HRu(CO)_4]^-$ and $[Ru(CO)_4I_2]$ react to give a formyl species. In other words, the first proposed step for the observed selectivity is an intermolecular reaction between $[HRu(CO)_4]^-$ and $Ru(CO)_4I_2$. Once the formyl species is formed, it is thought to undergo reactions similar to the ones shown in Fig. 4.7 to give oxygenated hydrocarbons.

4.5 RHODIUM-CATALYZED CARBONYLATION OF OTHER ALCOHOLS

Rhodium-catalyzed carbonylation of alcohols such as ethanol, n-propanol, and i-propanol have been studied. Unlike methanol carbonylation, central to all

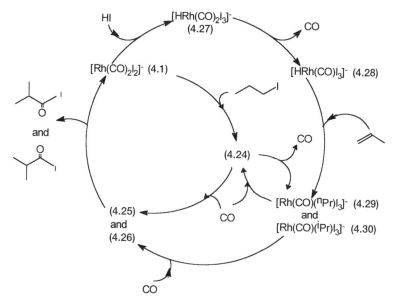

Figure 4.8   Hydrocarboxylation and carbonylation of propylene and *n*-propyl iodide. The outer cycle is the hydrocarboxylation pathway, while the inner cycle is the carbonylation pathway. The carbonylation pathway generates only *n*-butyric acid iodide. Note that 4.27 is identical with 4.5. A different designation is used to avoid cross-referencing.

these reactions is the reaction of alkenes with hydrido complexes that act as catalytic intermediates. Reactions where a carboxylic acid is formed from the reaction of an alkene with water and CO, as in reaction 4.9, is known as hydrocarboxylation (see Section 1.5). The outer catalytic cycle in Fig. 4.8

$$R\diagup\!\!=\ +\ CO\ +\ H_2O\ \longrightarrow\ R\diagup\!\!\diagdown\!\!\diagdown_{OH}^{O}\ +\ \diagup\!\!\diagdown\!\!_{R}^{O}{}_{OH}$$

(4.09)

shows the hydrocarboxylation route of propylene to *n*-butyric and *i*-butyric acid iodides. The inner cycle is the carbonylation route of *n*-propyl iodide to *n*-butyric acid iodide.

Although at a first glance the cycles may appear to be too complex, most of the catalytic species are analogues of intermediates shown in Fig. 4.2. Complexes 4.24 and 4.27 are oxidative addition products of 4.1 with *n*-propyl iodide and HI, respectively.

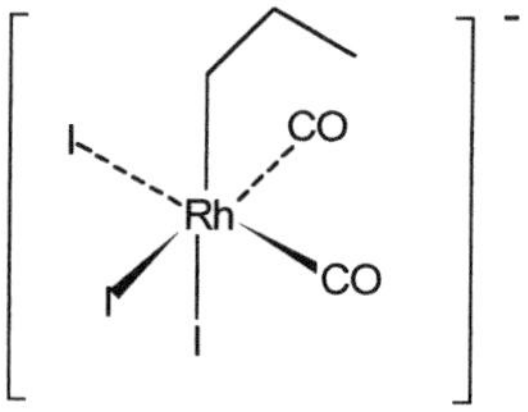

(4.24)

The latter complex undergoes CO loss to generate coordinatively unsaturated 4.28. Conversion of 4.28 to 4.30 is the crucial step that is responsible for the formation of the branched isomer. Obviously this reaction is possible only when propylene is present as one of the reactants, or under reaction conditions where propylene from *n*-propanol is generated in situ. Conversion of 4.28 to 4.30 is an example of alkene insertion into an M–H bond in a Markovnikov manner (see Section 5.2.2 for a discussion on Markovnikov and anti-Markovnikov insertion). The anti-Markovnikov path leads to the formation of 4.29, which is in equilibrium with 4.24. Complexes 4.25 and 4.26 are analogues of 4.4 with *n*-butyl and *i*-butyl groups in the place of methyl. They reductively eliminate the linear and branched acid iodides. In the presence of water the acid iodides are hydrolyzed to give *n*-butyric and *i*-butyric acids.

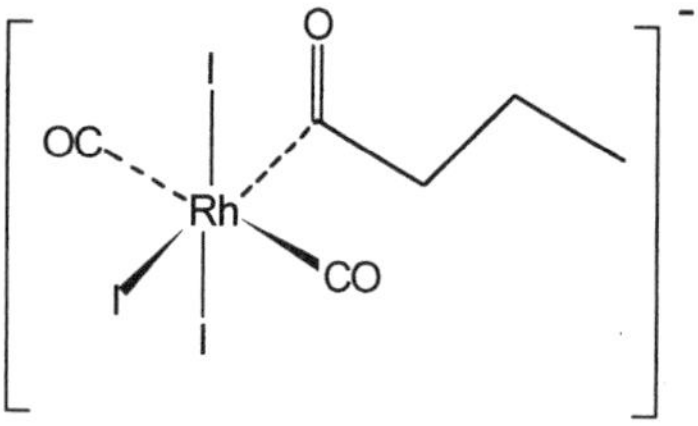

(4.25) 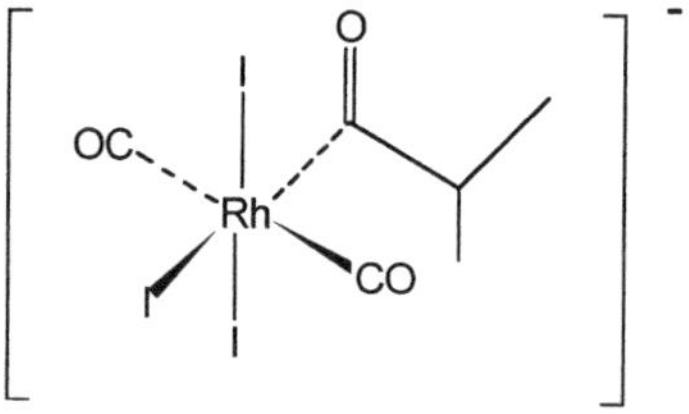　　　　　　　　　　　(4.26)

4.6　CARBONYLATION OF METHYL ACETATE

Conventional manufacture of acetic anhydride is based on the reaction of ketene ($H_2C\!=\!C\!=\!O$) with acetic acid. Eastman Chemical Company, a division of Eastman Kodak, set up facilities to manufacture 500 million lbs/yr of acetic anhydride and 165 million lbs/yr of acetic acid using coal as the feedstock in 1984. In this process coal is gasified to give synthesis gas, which is then converted to methanol by a heterogeneous catalytic process. Reaction of methanol with acetic acid gives methyl acetate, which is carbonylated to give acetic anhydride. Eastman Kodak consumes more than 1 billion lbs/yr of acetic anhydride in the production of cellulose esters used in the production of photographic films, plastics, coating chemicals, etc. The major factors behind the spectacular success of this process are as follows. All the production facilities are located in the heart of Appalachian coalfields. This process is less energy intensive than the ketene-based process. The carbonylation process recycles acetic acid, a coproduct in the manufacture of cellulose esters, in the synthesis of methyl acetate. Finally, conversions and rhodium metal recovery efficiencies of this process are sufficiently high to make the overall economics viable and attractive.

4.6.1 Mechanism and Catalytic Cycle

The acetic acid–forming part of the catalytic cycle for methanol carbonylation consists of reactions between acetyl iodide and water to give acetic acid and HI (Fig. 4.2, bottom left). The hydroiodic acid reacts with methanol to regenerate CH_3I and water. A similar mechanism operates for the carbonylation of methyl acetate. Acetic acid and acetyl iodide react to give acetic anhydride and HI. The latter reacts with methyl acetate to regenerate acetic acid and methyl iodide. These reactions are shown in Fig. 4.9 by the large, left-hand-side loop.

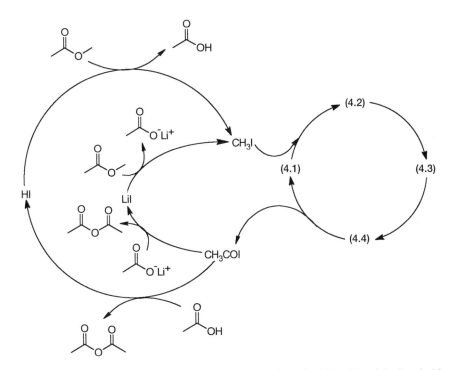

Figure 4.9 Carbonylation of methyl acetate to acetic anhydride. The right-hand-side cycle is the same as in 4.1 and 4.2. The big cycle on the left is similar to the organic cycle of 4.2, where acetic acid and methyl acetate substitute for water and methanol. The small inner cycle on the left is the dominant product-forming pathway.

In addition to this, there is another lithium salt promoted pathway (Fig. 4.9) that contributes significantly to product formation. Here the product-forming reaction between lithium acetate and acetyl iodide is followed by the reaction between LiI and methyl acetate. These reactions are shown by the inner loop on the left-hand side. In fact, the inner loop is the dominant product-forming pathway, and lithium salts play a crucial role in the overall catalysis. Note that the right-hand-side loop of the catalytic cycle is exactly the same as in Fig. 4.1(a).

Evidences for the proposed catalytic cycles come from kinetic and spectroscopic studies. In situ IR investigation indicates the presence of 4.1 as the main catalytic intermediate. The rate of acetic anhydride formation shows a complex dependence on the concentrations of rhodium, CH_3I, and lithium. Under conditions where lithium concentration is high, the rate is first order with respect to rhodium and CH_3I; that is, the rate is given by Eq. 4.3. However, with low lithium concentration the rate is independent (i.e., zero order) with respect to rhodium and methyl iodide concentrations. These observations are easily explained by identifying the slowest step under these two sets of conditions. With enough lithium the rates of reactions involving lithium salts are obviously high. Under these conditions oxidative addition of CH_3I to 4.1 is the slowest step, and rate expression 4.3 is obeyed. With low lithium concentration any one of the steps that involve lithium salts becomes the rate-determining one. The overall rate in such a situation is independent of rhodium and CH_3I concentrations.

4.7 CARBONYLATION OF ALKYNES; MANUFACTURE OF METHYL METHACRYLATE

The polymer of methyl methacrylate (MMA) is known as "Perspex." It is a clear transparent glasslike material with high hardness, resistance to fracture, and chemical stability. The conventional route, as shown by reaction 4.10, involves the reaction between acetone and hydrocyanic acid, followed by sequential hydrolysis, dehydration, and esterification. This process generates large quantities of solid wastes. An alternative route based on a homogeneous palladium catalyst has recently been developed by Shell. In this process a palladium complex catalyzes the reaction between propyne (methyl acetylene), methanol, and carbon monoxide. This is shown by reaction 4.11. The desired product is formed with a regioselectivity that could be as high as 99.95%.

(4.10)

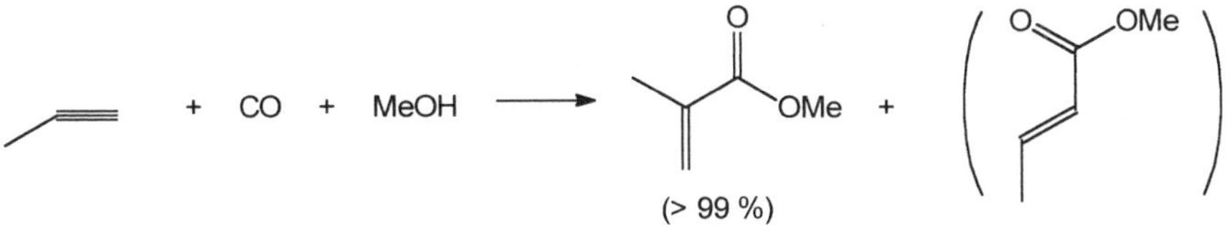

$$\text{HC} \equiv \text{C-CH}_3 \; + \; CO \; + \; MeOH \longrightarrow \underset{\text{(> 99 \%)}}{\overset{O}{\underset{\|}{\text{C}}}} \text{OMe} \; + \; \left(\overset{O}{\overset{\|}{\text{OMe}}} \right)$$

(4.11)

This invention has its roots in Reppe chemistry. In the late 1930s, Reppe in Germany had developed a number of manufacturing processes for bulk chemicals, where acetylene was used as one of the basic building blocks. Even today BASF and Rohm Hass manufacture large quantities of acrylic acid and its esters by hydrocarboxylation of acetylene. This reaction, 4.12, is catalyzed by a mixture of $NiBr_2$ and CuI. It involves high pressure (100 bar) and temperature (220°C), and mechanistically is not fully understood.

$$\text{HC} \equiv \text{CH} \; + \; CO \; + \; H_2O \longrightarrow \overset{O}{\overset{\|}{\text{C}}} \text{OH}$$

(4.12)

In contrast, the Shell process for MMA operates under milder conditions (60°C, 10–60 bar), and the mechanism at a molecular level is better understood. The reaction is usually carried out in methanol, which acts both as a solvent and as a reactant. The precatalyst is $Pd(OAc)_2$, which is mixed with an excess of phosphine ligand to generate the active catalytic intermediates in situ. An important requirement for efficient catalysis is the presence of an acid HX that acts as a co-catalyst.

4.7.1 Mechanism and Catalytic Cycle

The Shell process uses a novel ligand, 4.31, a phosphine that also has a pyridine ring. This ligand in the presence of an acid may act as a bidentate as well as a monodentate ligand. This is because the acid protonates the nitrogen atom of the pyridine ring and an equilibrium, as shown by 4.13 is set up.

$$\underset{(P \frown N)}{\overset{Ph}{\underset{}{\text{P}}} \overset{Ph}{}} \; + \; HX \; \rightleftharpoons \; \left[\underset{([P \frown NH]X)}{\overset{Ph}{\underset{}{\text{P}}} \overset{Ph}{}} \right] X^-$$

(4.13)

Under this condition both chelation, as shown by 4.32, and monodentate coordination, as shown by 4.33, may occur. Although not shown, in 4.33 the fourth coordination site is occupied by a solvent molecule or weakly coordinating X^-. A hypothetical catalytic cycle, based on these considerations, is shown in Fig. 4.10. In its protonated form ligand 4.31 acts as a labile, weakly coordinating ligand, easily displaced by reactants such as CO, methyl acetylene, etc.

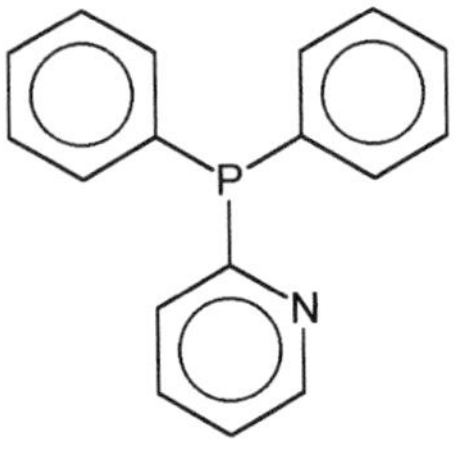

(4.31)

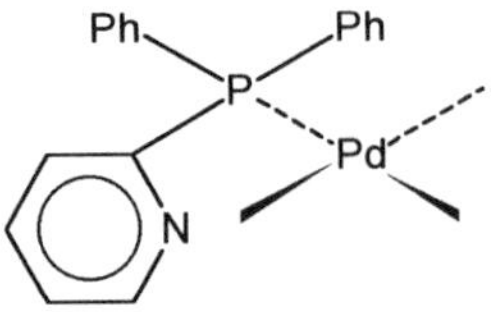

(4.32)

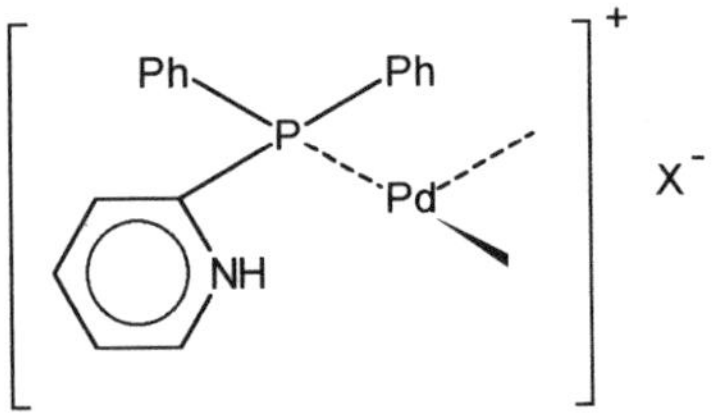

(4.33)

As shown in the proposed cycle, mixing $Pd(OAc)_2$ with excess ligand and a noncoordinating acid HX in methanol probably produces a species such as 4.34. Substitution of the protonated ligand by CO gives 4.35. Carbon monoxide insertion into the Pd–OMe bond and coordination by the protonated ligand produces 4.36. The ligand is displaced again by methyl acetylene to give 4.37. Insertion of the alkyne into the $Pd-CO_2Me$ bond, and coordination by the displaced ligand gives 4.38. A proton transfer from the protonated ligand to the σ-bonded alkene leads to the formation of MMA and 4.39. The latter reacts with methanol to complete the catalytic cycle and regenerate 4.34.

The catalytic cycle, though reasonable, is hypothetical. The proposed reactions and complexes have precedents in organometallic chemistry. The rate of the overall reaction is first order with respect to methyl acetylene and independent (zero order) of acid concentration as long as sufficient acid and 4.31 are present. Indirect but strong evidence for the proposed mechanism comes from structural modifications of the ligand and effects of such modifications

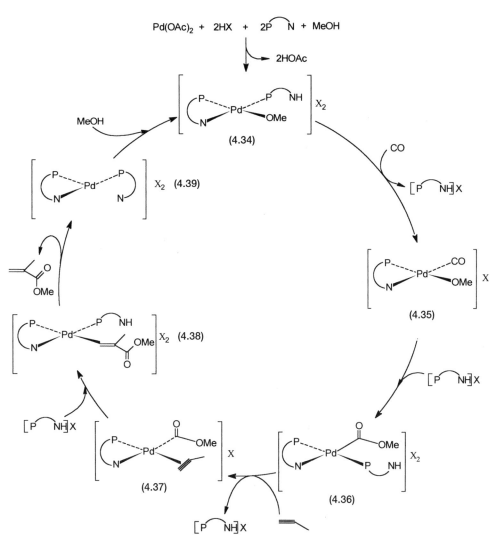

Figure 4.10 Shell process for the carbonylation of propyne (methyl acetylene) in methanol to give MMA. In 4.39 one P ∩ N may act as a monodentate ligand, and the fourth co-ordination site may be occupied by weakly coordinating X⁻.

on the rate. The rate of MMA formation is three orders of magnitude higher with 4.31 than that of analogues with a PPh₂ group in the 3- or 4-position of the pyridine ring. It is obvious that such ligands are incapable of acting as bidentate ligands. Molecular modeling results are also consistent with the proposed mechanism. However, no direct in situ spectroscopic evidence is available to support the mechanism.

4.8   OTHER CARBONYLATION AND HYDROCARBOXYLATION REACTIONS

There are some relatively small-volume but value-added chemicals that are commercially manufactured by carbonylation or hydrocarboxylation reactions. A few examples with some details are given in Table 4.2.

The mechanism of the Montedison reaction has been studied in some detail, and tentative mechanisms have been offered. The proposed catalytic cycle is shown in Fig. 4.11. The biphasic reaction medium consists of a layer of diphenyl ether and that of aqueous alkali. In the presence of alkali, the precatalyst $Co_2(CO)_8$ is converted into 4.7. The sodium salt of 4.7 is soluble in water but can be transported to the organic phase, that is, a diphenyl ether layer by a phase-transfer catalyst. The phase-transfer catalyst is a quaternary ammonium salt ($R_4N^+X^-$). The quaternary ammonium cation forms an ion pair with $[Co(CO)_4]^-$. Because of the presence of the R groups, this ion pair, $[R_4N]^+[Co(CO)_4]^-$, is soluble in the organic medium. In the nonaqueous phase benzyl chloride undergoes nucleophilic attack by 4.7 to give 4.40, which on carbonylation produces 4.41. The latter in turn is attacked by hydroxide ion transported from the aqueous phase, to the organic phase again by the phase-transfer catalyst. The product phenyl acetate and 4.7 are released in the aqueous phase as the sodium or quaternary ammonium salts.

As mentioned in Table 4.2, the industrial Ube reaction is probably carried out on a heterogeneous catalyst. The probable mechanism is based on generation of an alkyl nitrite and oxidative addition of the alkyl nitrite onto a Pd^0 center. Nitric oxide is used as a co-catalyst in the Ube process. The Hoechst

TABLE 4.2   Chemicals Manufactured by Carbonylation or Hydrocarboxylation Reactions

Manufacturer	Product	Process
Montedison	Phenylacetic acid; an intermediate for pesticide and perfumes	Cobalt catalyst; $PhCH_2Cl$, CO, and HO^- are the reactants; $T \sim 55°C$, P = a few bars; Reaction carried out in a biphasic medium with a *phase-transfer* catalyst
Ube Industries	Oxalate diesters	Palladium catalyst; ROH, CO, and O_2 are reactants. Nitric oxide is used as a co-catalyst; Actual production is probably based on a heterogeneous catalyst
Hoechst	Ibuprofen; an analgesic (see Table 1.1)	Carbonylation of appropriate secondary alcohol with a palladium catalyst

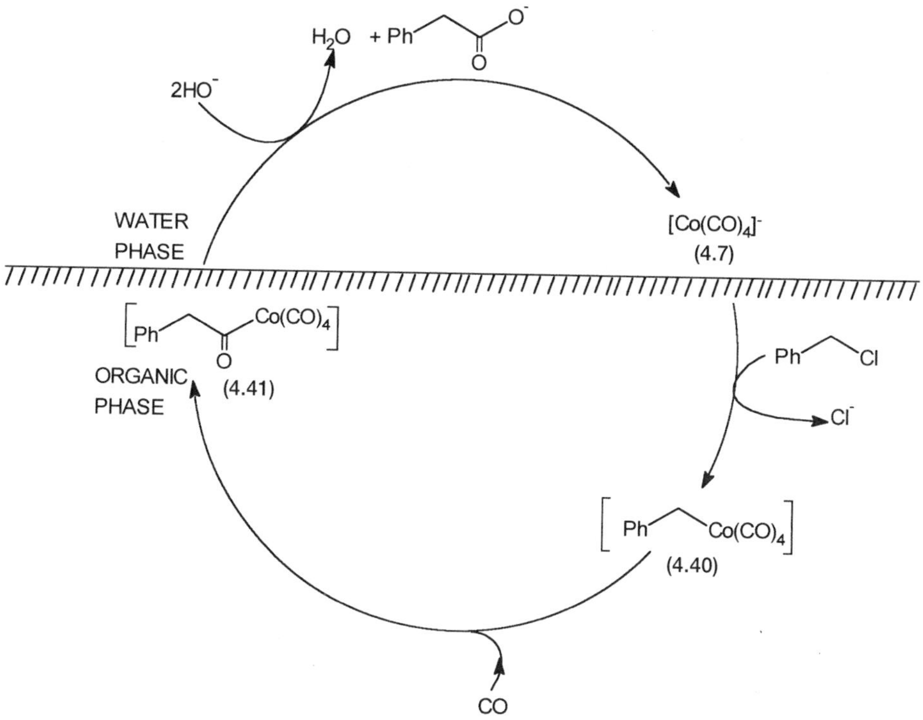

The dashed line, ⊓⊓⊓⊓⊓⊓⊓, represents the phase boundary in the biphasic system.

Figure 4.11   The Montedison process for the carbonylation of benzyl chloride in a biphasic system. A phase-transfer catalyst is used to facilitate transport of $Co(CO)_4^-$ and HO^- from the aqueous to the organic phase.

Celanese process for the manufacture of ibuprofen on a 3500-ton scale has been operating since 1992. In this process isobutyl benzene is acylated and then hydrogenated over a heterogeneous catalyst to give the appropriate precursor alcohol. This alcohol is then carbonylated. The overall synthetic scheme is shown by reaction 4.14. The conventional process for ibuprofen manufacture was based on six synthetic steps and generated a large amount of salt as a solid waste.

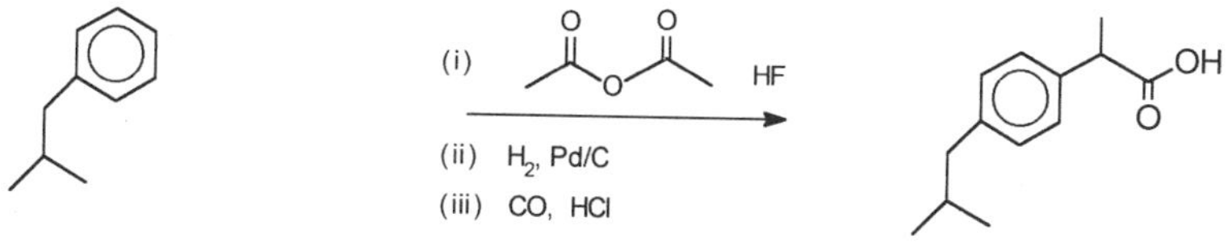

(4.14)

The carbonylation reaction in the Hoechst process involves the use of PdCl$_2$(PPh$_3$)$_2$ as the precatalyst, a CO pressure of about 50 bar, and a temperature of about 130°C. It is performed in a mixture of an organic solvent and hydrochloric acid. The mechanism at a molecular level is not known with certainty. On the basis of the known chemistry of palladium, a speculative catalytic cycle is shown in Fig. 4.12.

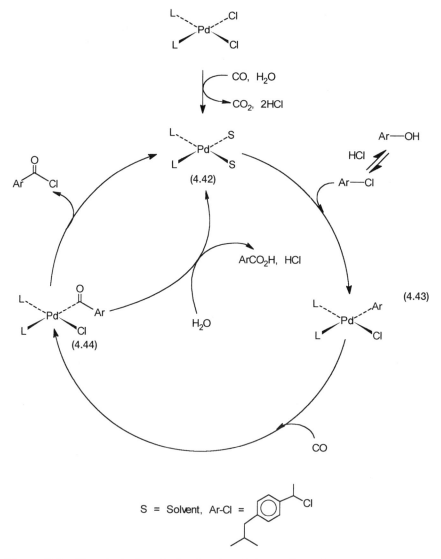

Figure 4.12 Hypothetical cycle for Hoechst–Celanase ibuprofen process. For all the proposed intermediates *cis* geometry is assumed.

The basic assumptions are as follows: The precatalyst is first reduced to the zero-valent palladium species 4.42. Oxidative addition of Ar–Cl to 4.42 gives 4.43. CO insertion into the Pd–C bond results in the conversion of 4.43 to 4.44. Finally, ArCOCl is reductively eliminated from 4.44. The acid chloride undergoes hydrolysis to give ibuprofen. Direct reaction between 4.44 and water to give 4.42, ibuprofen, and HCl is also possible.

4.9 ENGINEERING ASPECTS

A block diagram of the Monsanto process for acetic acid production is shown in Fig. 4.13. The process flow sheet is simple since the reaction conditions are mild (180°C/30–40 bar) when compared to the BASF process (250°C/700 bar). More than 40% of world's acetic acid is made by the Monsanto process. One of the problems with this process is the continuous loss of iodine. A block diagram of the Eastman process for acetic anhydride production is shown in Fig. 4.14. The process generates minimum waste, and all process tars are destroyed to recover iodine and rhodium.

Carbon monoxide free from impurities is required for the carbonylation reactions. The selection of the purification process depends on several factors, including the level of impurity permitted and the source of gas that needs to be purified. Carbon monoxide content in the source stream generally varies between 6 and 60%, while the carbon dioxide content is between 3 and 15%. Blast furnace gas will have predominantly nitrogen (of the order of 60%), while coke oven, steam reforming, and water gases will have hydrogen in the range of 30–75%. Coke oven and water gases will also have methane as an additional component (12–28%).

Carbon monoxide is first stripped off particulate matter in a cyclone separator or in a scrubber. Scrubbing also removes tar or heavy hydrocarbon fractions. Acidic gases, if present, are removed by absorption in monoethanol amine or in potassium carbonate. This pretreated gas is sent to the next section for further purification. Commercial processes for final purification are based on the absorption of carbon monoxide by salt solution, low-temperature condensation, or fractionation, or by pressure-swing adsorption using a solid material.

The salt process employs scrubbing of the gas with cuprous ammonium salt solution. Carbon monoxide forms a complex with the solution at high pressure and low temperature in the absorption column. The absorbed pure carbon monoxide is released from the solution at low pressure and high temperature in the regenerator/stripper section. Any carbon monoxide that is liberated in the stripper section is removed by subsequently washing the gas with caustic. Sulfur has to be removed in the pretreatment stage to prevent the formation of solid sulfide.

In the Cosorb process, cuprous tetrachloroaluminate toluene complex in a toluene medium is used as the absorbing liquid instead of the copper salt so-

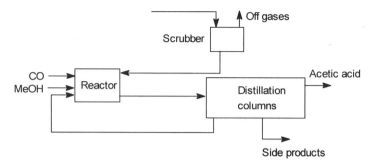

Figure 4.13   Simplified block diagram of the Monsanto process for the acetic acid production.

lution. A low corrosion rate and low energy consumption are the advantages of the Cosorb process over the previous one.

In the cryogenic partial condensation process, the feed gas is cooled to 85 K, which condenses the bulk of the carbon monoxide and methane. Uncondensed hydrogen-rich gas is cooled further to 70 K to liquefy additional carbon monoxide. These liquid fractions are degassed and then warmed to obtain high-purity carbon monoxide gas. A cryogenic distillation process can produce ultrapure carbon monoxide.

High-purity carbon monoxide (>99.8%) can be obtained in the liquid methane wash process, where the inlet gas is cooled to about 90 K and the bulk of the carbon monoxide is removed by condensation. The hydrogen-rich stream is washed in a column with methane, which absorbs carbon monoxide and other gases. Carbon monoxide is recovered from this solution in a fractional distillation column.

Pressure swing adsorption (PSA) utilizing copper adsorbent is used to separate carbon monoxide from blast furnace and coke oven gases. The copper adsorbent is finely dispersed on a high-surface-area support such as alumina or carbon. During the adsorption cycle carbon monoxide is selectively adsorbed

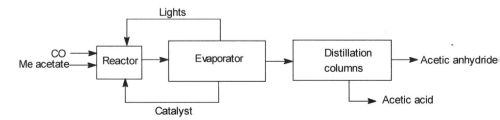

Figure 4.14   Simplified block diagram of the methyl acetate carbonylation process for acetic anhydride production.

from the gas mixture. During the depressurization cycle carbon monoxide is desorbed from the solid matrix.

PROBLEMS

1. Identical activities are observed for methanol carbonylation when $RhI(CO)(PR_3)_2$ (R=Ph, Et, Me) are used as catalysts. Why?

Ans. All get converted to 4.1 to start the catalytic cycle.

2. In Fig. 4.1 if the step 4.4 to 4.1 were rate-determining, what kind of rate expression would have been expected?

Ans. Rate should have shown dependence on CO pressure.

3. From the discussion in Section 4.2.2., which step of Fig. 4.1 must be reversible? In a free-energy diagram what would be the relative placement of 4.2 and to 4.3?

Ans. 4.1 to 4.2. 4.1 and 4.3 considerably lower than 4.2; activation energy barriers for 4.2 to 4.1 and 4.2 to 4.3 are comparable and small (see the reference in answer to Question 4).

4. Use of neat CH_3I as the solvent allows spectroscopic identifications of 4.2, 4.3, and 4.4 but not if CH_3I is diluted with other solvents. Why?

Ans. Sufficient steady-state concentration of 4.2 is required for spectroscopic identification (see P. M. Maitlis et al., *J. Chem. Soc. Dalton Trans.* 2187– 96, 1996; A. Haynes et al., *J. Am. Chem. Soc.* **115**, 4093–100, 1993).

5. Assume that the basic mechanism of the Monsanto process is valid for the cobalt-based process. What would be the crucial hypothetical catalytic intermediates?

Ans. $[Co(CO)_3(CH_3)I]^-$, $[Co(COCH_3)(CO)_2)I]^-$, etc.

6. What products are expected in the hydrocarboxylation of (a) 1-butene and (b) 2-butene?

Ans. (a) 1-pentenoic and 2-methyl butyric acid; (b) 2-methyl butyric acid.

7. The rate of formation of isobutyric acid in the carbonylation of *i*-propanol is strongly dependent on pH, but carbonylation of *n*-propanol to give the same product is much less sensitive to pH. Why?

Ans. Hydrocarboxylation is the main route for *i*-propanol carbonylation (under the reaction conditions propylene is generated). Low pH favors formation of 4.27.

8. Theoretically how many isomers are possible for the catalytic intermediate 4.3? In ^{13}C NMR and carbonyl IR, a sharp and a broad signal are, respectively, observed. What does this indicate?

Ans. 3 square pyramids, 4 trigonal bypyramids; the isomers equilibrate at rates faster than that of the NMR time scale but slower than that of the IR time scale.

9. Do you expect differences between the rates of carbonylation of (a) $CD_3CDODCD_3$ and $CH_3CHOHCH_3$ by the Monsanto process? (b) CH_3OH and C_2H_5OH by the Monsanto and BASF processes? (c) CD_3OD and CH_3OH by the Monsanto process? If so, why?

Ans. (a) Yes; hydrocarboxylation mechanism, C–H bond breaking precedes the rate-determining step. (b) Slight differences reflecting the difference in the rate of nucleophilic substitutions of CH_3I and C_2H_5I. (c) No; C–H bond cleavage not involved in mechanism.

10. How would the byproduct formation in the rhodium-catalyzed carbonylation reaction be affected by the absence of CH_3I in an acidic pH?

Ans. No acetic acid cycle. Enhancement of water-gas shift reaction.

11. In the BASF process, the presence of hydrogen has an adverse effect, while in the Monsanto process small amounts of hydrogen increase efficiency. Why?

Ans. In the BASF process more Fischer–Tropsch by-products are formed. In "Rh"-based catalysis rhodium is maintained in the lower oxidation state, ensuring that 4.1 concentration is higher than 4.12 and the acetic acid–forming cycle dominates.

12. The ^1H NMR spectra of many metal carbonyls after treatment with $NaBH_4$ show a signal at $\delta \cong 10.6$ ppm. What relevance does this information have to carbonylation reaction by cobalt?

Ans. The signal is due to the coordinated formyl group, a crucial hypothetical step in cobalt-catalyzed Fischer–Tropsch reactions that produce oxygenated hydrocarbons.

13. In methyl acetate carbonylation the plot of rate against the concentration of lithium iodide has a small positive intercept. Do you expect it to be the same for two different rhodium ion concentrations?

Ans. In the absence of lithium iodide ($[Li^+] = 0$), the bigger left-hand cycle operates. Under these conditions, if the slowest step were the reaction of acetyl iodide with acetic acid or HI with methyl acetate, then the rate would be independent of rhodium concentration. If oxidative addition of CH_3I on 4.1 is the slowest step, then the intercepts will be different for different rhodium concentrations.

14. ^{13}CO-labeled (4.1) when reacted with HI at $-80°C$ shows $\delta(^1H)$ at -8.40 ppm in ^1H NMR and $\delta(^{13}C)$ at 173.83 ppm in ^{13}C NMR. On warming the solution to $-20°C$, $\delta(^1H)$ at -10.45 and $\delta(^{13}C)$ at 177.9 ppm are observed.

Explain these observations and their relevance to ethanol carbonylation at low pH.

Ans. Formation of 4.27 (*trans* COs) at −20°C. At −100°C a five-coordinate neutral I⁻ dissociated intermediate (see D. C. Roe et al., *J. Am. Chem. Soc.* **116**, 1163–64, 1994). Spectroscopic evidence for the hydrocarboxylation cycle where 4.27 is a crucial intermediate.

15. In the Shell process for MMA manufacture an alternative mechanism involving a hydride intermediate may be invoked. Draw a catalytic cycle based on such a mechanism.

Ans. Insertion of propyne into Pd–H in a Markovnikov manner followed by CO insertion into the Pd–C bond followed by methanolysis.

16. In the Shell process methyl substitution in the pyridine ring of 4.31 at the 2, 3, and 4 positions results in maximum selectivity for the 2-substituted ligand. The activities in the three cases are more or less the same. What mechanistic conclusion may be drawn from this?

Ans. The mechanism of Question 15 does not operate, since steric hindrance should favor anti-Markovnikov addition and lowering of selectivity.

17. The −20°C species of Question 14 undergoes slow disproportionation to 4.1, 4.13, and dihydrogen at 0°C. Based on this observation, how can the catalytic cycle of Fig. 4.5 be modified?

Ans. From 4.1 to 4.12 involvement of 4.27 and the I⁻ dissociated species as transitory intermediates.

18. Show how under carbonylation conditions 4.24 may lead to a mixture of 4.25 and 4.26, and finally to a mixture of normal and isobutyric acid from *n*-propanol.

Ans. 4.24 undergoes β-elimination to give 4.27 and propylene. Insertion of propylene into the Rh–H bond in the Markovnikov and anti-Markovnikov manner followed by CO insertion gives 4.26 and 4.25. In other words, reversible β-elimination and insertion initiates the hydrocarboxylation cycle.

19. Complexes ML_4 (M = Pt or Pd, L = PPh_3) have recently been used as catalysts for the reactions between RC≡CH, R′SH, and CO. Are these reactions similar to any discussed in this chapter? What are the possible products and mechanisms for their formation?

Ans. Yes, similar to the Shell reaction for MMA (see *J. Am. Chem. Soc.* **119**, 12380–81, 1997).

BIBLIOGRAPHY

For All the Sections

Books

Most of the books listed under Sections 1.3 and 2.1–2.3.4 contain information on carbonylation reactions and should be consulted. Especially useful are the book by Parshall and Ittel and Section 2.1.2 of Vol. 1 of *Applied Homogeneous Catalysis with Organometallic Compounds*, ed. by B. Cornils and W. A. Herrmann, VCH, Weinheim, New York, 1996.

For palladium-based carbonylation, see *The Organic Chemistry of Palladium*, P. M. Maitlis, Academic Press, New York, 1971.

For Section 4.9 see Kirk and Othmer, *Encyclopedia of Chemical Technology*, 4th edn, John Wiley & Sons, New York, Vol. 4, 1992, pp. 798–809; Ullman's *Encyclopedia of Industrial Chemistry*, 5th edn, VCH, New York, 1992, Vol. A1, pp. 47–55.

Articles

Sections 4.1 to 4.2

R. P. A. Sneeden, in *Comprehensive Organometallic Chemistry*, ed. by G. Wilkinson, F. G. A. Stone, and E. W. Abel, Pergamon Press, Vol. 8, 1982, pp. 19–100.

T. W. Dekleva and D. Forster, in *Advances in Catalysis*, ed. by D. D. Eley, H. Pines, and P. B. Weisz, Academic Press, New York, Vol. 34, 1986, pp. 81–130.

D. Forster, in *Advances in Organometallic Chemistry*, ed. by F. G. A. Stone and R. West, Academic Press, Vol. 17, 1979, pp. 255–68.

D. Forster and T. W. Dekleva, *J. Chem. Edu.* **63**, 204–6, 1986.

A. G. Kent et al., *J. Chem. Soc. Chem. Commun.*, 728–29, 1985.

A. Haynes et al., *J. Am. Chem. Soc.* **115**, 4093–100, 1993.

For a recent lucid summary, see P. M. Maitlis et al., *J. Chem. Soc. Dalton Trans.*, 2187–96, 1996.

Section 4.3

P. C. Ford and A. Rokicki, in *Advances in Organometallic Chemistry*, ed. by F. G. A. Stone and R. West, Vol. 28, pp. 139–217.

Section 4.4

C. Masters, in *Advances in Organometallic Chemistry*, ed. by F. G. A. Stone and R. West, Academic Press, 1979, pp. 61–104.

R. P. A. Sneeden, in *Comprehensive Organometallic Chemistry*, Ed. by G. Wilkinson, F. G. A. Stone, and E. W. Abel, Pergamon Press, Vol. 8, pp. 19–100, 1982.

B. D. Dombek, in *Advances in Catalysis*, Ed. by D. D. Eley, H. Pines, and P. B. Weisz, 1983, Academic Press, New York, Vol. 32, pp. 326–416.

B. D. Dombek, *J. Chem. Edu.* **63**, 210–12, 1986.

J. A. Gladysz, in *Advances in Organometallic Chemistry*, ed. by F. G. A. Stone and R. West, Vol. 20, Academic Press, pp. 1–38.

For 4.23, see D. Milstein et al., *J. Am. Chem. Soc.* **108**, 3711–18, 1986.

Section 4.5

See references listed under 4.1 and 4.2, especially the articles by D. Forster.

Section 4.6

S. W. Polichnowski, *J. Chem. Edu.* **63**, 206–9, 1986.

V. H. Agreda et al., *Chemtech* **22**(3), 172–181, 1992; ibid. **18**(4), 250–53, 1988.

Section 4.7

E. Drent et al., in *Applied Homogeneous Catalysis with Organometallic Compounds* (see under Books), Vol. 2, pp. 1119–31.

Also see E. Drent et al., *J. Organomet. Chem.* **475**, 57–63, 1994; ibid. **455**, 247–53, 1993.

Section 4.8

M. Beller, in *Applied Homogeneous Catalysis with Organometallic Compounds* (see under Books), Vol. 1, pp. 148–58.

For phenyl acetic acid, see L. Casar, *Chem. Ind. (Milan)* **67**, 256, 1985.

For ibuprofen, see J. N. Armor, *Appl. Catal.* **78**, 141, 1991; *J. Mol. Catal. A-Chem.* **138**(1), 25–26, 1999.

CHAPTER 5

HYDROFORMYLATION

5.1 BACKGROUND

Otto Roelen at Ruhrchemie AG discovered the hydroformylation or oxo reaction in 1938. As shown by reaction 5.1, the basic reaction is the addition of a hydrogen atom and a formyl group to the double bond of an alkene. The reaction works efficiently, mainly with terminal alkenes. With an optimal choice of ligands and process conditions, very high selectivity (>95%) for the desired isomer of the aldehyde could be achieved.

$$(5.1)$$

The aldehydes commercially produced this way are many. One of the most important is *n*-butyraldehyde. Isononyl aldehyde is also an important intermediate. As shown by reactions 5.2 and 5.3, propylene is hydroformylated to *n*-butyraldehyde which is then converted by aldol condensation and hydrogenation to 2-ethyl hexanol.

$$(5.2)$$

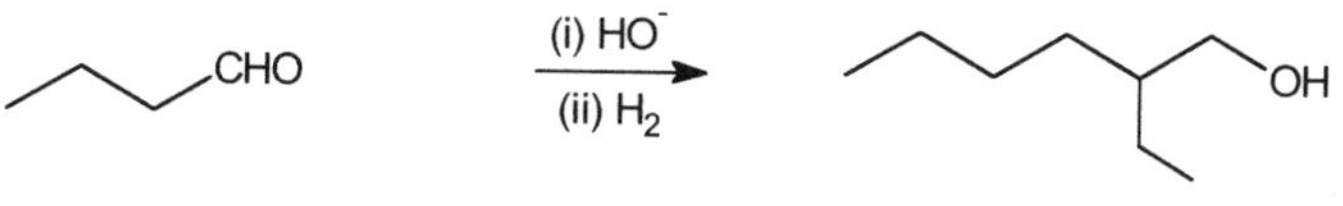

$$(5.3)$$

Both 2-ethyl hexanol and isononyl alcohol are used, in combination with phthalic anhydride, as plasticizers for polyvinyl chloride resin. Hydroformylation is also used for the manufacture of long-chain fatty alcohols. The linear alcohols are used in detergents and are more biodegradable than the branched ones. The Shell higher olefin process (SHOP) uses ethylene that is oligomerized, isomerized, methathesized, and then hydroformylated to give long-chain fatty alcohols (see Section 7.4.1).

Three commercial homogeneous catalytic processes for the hydroformylation reaction deserve a comparative study. Two of these involve the use of cobalt complexes as catalysts. In the old process a cobalt salt was used. In the modified current version, a cobalt salt plus a tertiary phosphine are used as the catalyst precursors. The third process uses a rhodium salt with a tertiary phosphine as the catalyst precursor. Ruhrchemie/Rhone-Poulenc, Mitsubishi-Kasei, Union Carbide, and Celanese use the rhodium-based hydroformylation process. The phosphine-modified cobalt-based system was developed by Shell specifically for linear alcohol synthesis (see Section 7.4.1). The old unmodified cobalt process is of interest mainly for comparison. Some of the process parameters are compared in Table 5.1.

5.2   THE RHODIUM PROCESS

The high selectivity and mild conditions make the rhodium process more attractive than the cobalt one for the manufacture of *n*-butyraldehyde. The high cost of rhodium makes near-complete catalyst recovery a must for the commercial viability of the process. As we shall see, this has been achieved by developing an elegant separation method based on water-soluble phosphines.

5.2.1   The Catalytic Cycle

The catalytic cycle and the catalytic intermediates for the rhodium-plus-phosphine-based process are shown in Fig. 5.1. It is important to note that hydroformylation with rhodium can also be effected in the absence of phosphine. In such a situation CO acts as the main ligand (i.e., in Fig. 5.1, L = CO). The mechanistic implications of this is discussed later (Section 5.2.4).

The following points are to be noted. First of all, complexes 5.1, 5.3, 5.5, and 5.7 are 18-electron complexes, while the rest are 16-electron ones. Second, conversions of 5.3 to 5.4 and 5.5 to 5.6 are the two insertion steps. The selectivity towards *n*-butyraldehyde is determined in the conversion of 5.3 to 5.4. It is possible that a rhodium–isopropyl rather than rhodium–propyl complex is formed. In such a situation on completion of the catalytic cycle isobutyraldehyde will be the product. In practice both the *n*-propyl and the *i*-propyl complexes of rhodium are formed, and a mixture of *n*-butyraldehyde and *i*-butyraldehyde is obtained. This aspect is discussed in greater detail in the following section. Third, the catalyst precursor 5.1 undergoes ligand dissocia-

TABLE 5.1   Process Parameters for Several Hydroformylation Processes

Process parameters	Cobalt	Cobalt + phosphine	Rhodium + phosphine
Temperature (°C)	140–180	160–200	90–110
Pressure (atm)	200–300	50–100	10–20
Alkane formation	Low	Considerable	Low
Main product	Aldehyde	Alcohol	Aldehyde
Selectivity (%) to n-butyraldehyde	75–80	85–90	92–95
Isolation of catalyst	Difficult; HCo(CO)₄ is volatile	Less difficult	Less difficult; water-soluble phosphine a major advancement

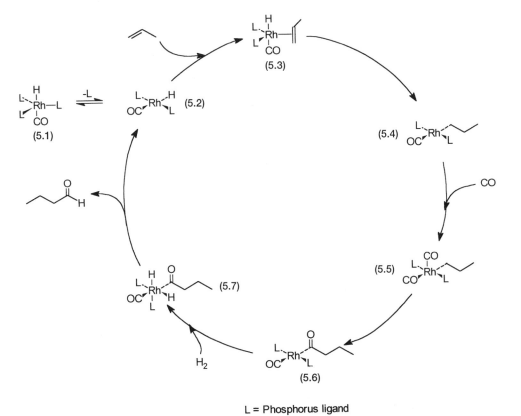

L = Phosphorus ligand

Figure 5.1 The basic catalytic cycle for the hydroformylation of propylene with Rh/PPh₃-based catalyst. In step 5.3 to 5.4 anti-Markovnikov addition is assumed.

tion to generate 5.2, a coordinatively unsaturated species. Finally, conversion of 5.6 to 5.7 is an oxidative addition reaction, while the conversion of 5.7 to 5.2 occurs by reductive elimination.

5.2.2 Product Selectivity

The two different ways of inserting an alkene into a metal–hydrogen bond, as shown by 5.4 and 5.5, are called anti-Markovnikov and Markovnikov addition, respectively. Insofar as hydroformylation with high selectivity to *n*-butyraldehyde is concerned, it is considered to be primarily an effect of steric crowding around the metal center. The normal alkyl requires less space and therefore formed more easily than the branched one in the presence of bulky ligands.

$$(5.4)$$

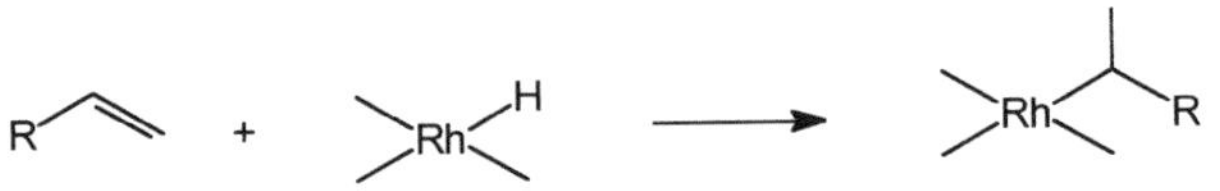

$$(5.5)$$

However, the balance between sterically demanding ligands and their ability to remain coordinated so that the product selectivity could be influenced is a fine one. This aspect is discussed in more detail in Section 5.2.4. Although not directly related to hydroformylation, it is appropriate to note here that Markovnikov additions accompanied by β-hydride elimination is a general pathway for alkene isomerization. This is shown in Fig. 5.2 for the isomerization of both terminal and internal alkenes.

5.2.3 Mechanistic Studies

Kinetic studies are of limited value for elucidating the mechanism of the hydroformylation reaction. This is because the empirically derived rate expressions are valid only within a narrow range of experimental conditions. For the rhodium-catalyzed reaction, in the *absence* of phosphine, the following rate expression has been proposed:

$$\frac{d[\text{RCHO}]}{dt} = k[\text{Rh}]^a[\text{alkene}]^b[\text{H}_2]/[\text{CO}] \qquad (5.6)$$

The value of b depends on the alkene, and both a and b under certain conditions may be less than one. The inverse dependence of rate on the concentration

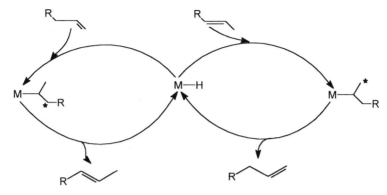

Figure 5.2   Alkene isomerization by alkene insertion into the M–H bond followed by β-hydride elimination. Note that the insertion is in a Markovnikov manner. The β-carbons from which β-hydride elimination leads to isomerization are marked by asterisks.

(pressure) of CO is observed only at total pressure exceeding 40 atm. At lower pressures the rate increases with increased pressure.

Industrially a large excess of the phosphine is used. This is necessary for good selectivity to n-butyraldehyde (see Section 5.2.4). On the basis of the mechanism proposed in Fig. 5.1, excess phosphine should reduce the reaction rate. This is because the precatalyst 5.1 must first undergo phosphine dissociation to initiate the catalytic cycle. The oxidative addition of dihydrogen to 5.6 is most likely the rate-determining step. Note that if we ignore the rate dependence on the phosphorus ligand, then the type of rate expression given by 5.6 is consistent with such a rate-determining step. The rate expression does indicate a transition state with interactions between rhodium and alkene, as well as rhodium and hydrogen.

Both in situ infrared and multinuclear NMR under less severe conditions have been used to gain mechanistic insights. For the hydroformylation of 3,3-dimethyl but-1-ene, the formation and hydrogenolysis of the acylrhodium species $Rh(COR)(CO)_4(R=CH_2CH_2Bu')$ can be clearly seen by IR. NMR spectroscopy has also been very useful in the characterization of species that are very similar to the proposed catalytic intermediates. We have already seen (Section 2.3.3, Fig. 2.7) NMR evidence for equilibrium between a rhodium alkyl and the corresponding hydrido–alkene complex. There are many other similar examples. Conversion of 5.3 to 5.4 is therefore well precedented. In the absence of dihydrogen allowing CO and alkene to react with 5.1, CO adducts of species like 5.6 can be seen by NMR. Structures 5.11 and 5.12 are two examples where the alkenes used are 1-octene and styrene, respectively.

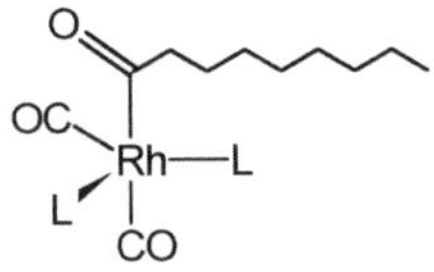

(5.11)

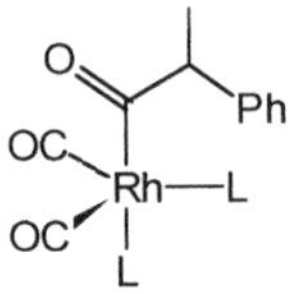

(5.12)

The other isomer of 5.11 with one axial and one equatorial phosphine can also be seen at low temperature. Indeed at low temperature that appears to be the more stable isomer. With styrene, as shown by 5.12 the branched (Markovnikov) rather than the linear (anti-Markovnikov) isomer is the major one. However, remember that the experimental conditions of an actual industrial process, and that of the NMR experiments, are different.

Model compounds for some of the catalytic intermediates shown in Fig. 5.1 have also been synthesized and characterized. Thus $trans$-Rh(C$_2$F$_4$H)(CO)L$_2$ is known, which may be considered as a model of 5.4. A model for 5.7, the complex 5.13, has also been made and characterized. While there are clear differences between 5.7 and 5.13, both 5.7 and 5.13 undergo reductive elimination. For the latter, the product of reductive elimination is acetaldehyde. This reaction has been kinetically studied to gain mechanistic insights.

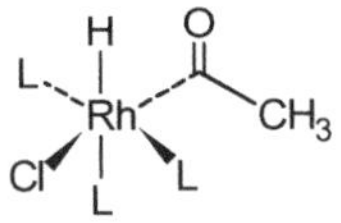

(5.13)

5.2.4   The Phosphorus Ligands and Selectivity

As has already been mentioned (Section 5.2.1), the catalyst precursor RhH(CO)L$_3$ has to undergo ligand dissociation to generate an unsaturated, catalytically active intermediate. In the absence of phosphines CO is the main ligand of the various catalytically active intermediates. It is obvious, therefore, that in the presence of both CO and L there are several possible equilibria. Under mild conditions, the ones shown in Fig. 5.3 have been observed by multinuclear (^{1}H, ^{13}C, ^{31}P) temperature-variable NMR.

In a typical industrial hydroformylation process, the rhodium to phosphorus molar ratio is between 1:50 to 1:100, while the partial pressure of CO is about

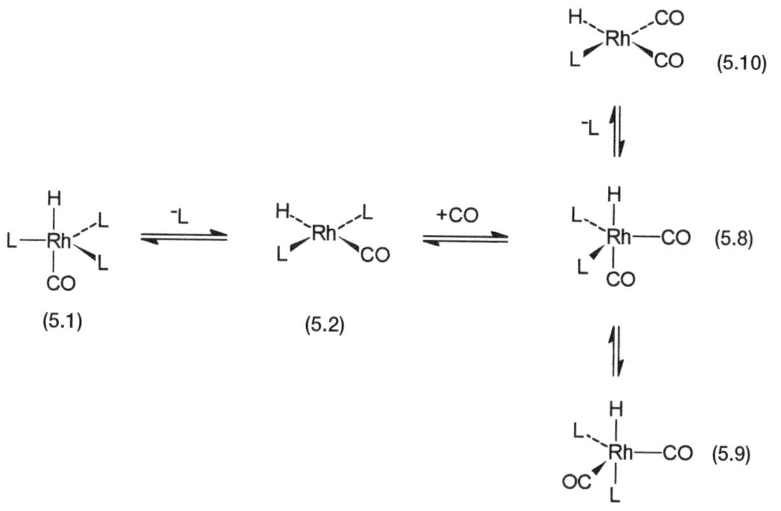

Figure 5.3   Equilibria involving dissociation of L (L = PPh$_3$). Note that 5.8 to 5.9 is an equilibrium that interchanges equatorial and axial ligands by Berry pseudorotation.

25 bar. It is reasonable to assume that under these conditions all the complexes with the general formula HRh(CO)$_n$L$_{4-n}$ ($n = 0$–4) might be present with variable concentrations. Like HRh(CO)L$_3$, the catalyst precursor in Fig. 5.1, all these species can undergo ligand dissociation and initiate catalytic cycles similar to the one already discussed. Indeed there are commercial hydroformylation processes where rhodium precatalysts are used without any phosphorous ligand (see Section 5.4). In such processes HRh(CO)$_4$-derived catalytic intermediates take part in catalysis.

The individual catalytic cycles generated by HRh(CO)$_n$L$_{4-n}$ species will be coupled with each other by ligand substitution reactions. This is shown in Fig. 5.4, where only the possible equilibria between the catalyst precursors (i.e., analogues of 5.1), the dissociated species (i.e., analogues of 5.2), and the alkene coordinated species (i.e., analogues of 5.3) are shown

It is obvious that such equilibria would exist for all the other catalytic intermediates. The result of all this is coupled catalytic cycles and many simultaneous catalytic reactions. This is shown schematically in Fig. 5.5. The complicated rate expressions of hydroformylation reactions are due to the occurrence of many reactions at the same time. As indicated in Fig. 5.5, selectivity towards anti-Markovnikov product increases with more phosphinated intermediates, whereas more carbonylation shifts the selectivity towards Markovnikov product. This is to be expected in view of the fact that a sterically crowded environment around the metal center favors anti-Markovnikov addition (see Section 5.2.2).

The use of bulky phosphorus ligands does not necessarily push the reaction towards the more phosphinated cycles. This is because too much steric crowd-

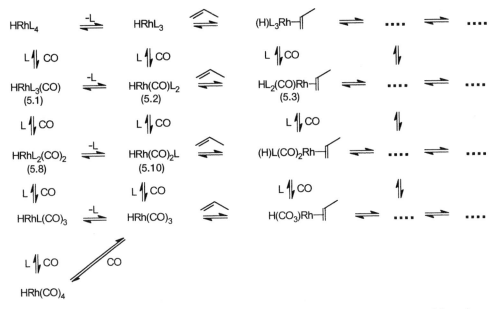

Figure 5.4   Some of the expected equilibria in the presence of large excess of L and CO. Only the analogues of 5.1, 5.2, and 5.3 are shown. The catalytic intermediates of the succeeding steps will occupy the vacant positions indicated by filled squares. Note that there is direct evidence for equilibrium between 5.1, 5.8, 5.2, and 5.10 from NMR.

ing around the metal center leads to phosphine dissociation. In such a situation the reaction in fact would be pushed towards cycles with less phosphinated intermediates, that is, towards Markovnikov addition. The above reasoning has been tested by using phosphorus ligands with different cone angles (see Section 2.1.2). Thus with the very bulky $P(OC_6H_3\text{-}2,6\text{-}Me_2)_3$, which has a cone angle $>170°$, the selectivity towards n-butyraldehyde is only about 50%. In contrast, $P(OC_6H_5)_3$, which has a cone angle $\sim 128°$, gives a selectivity of about 85%.

In summary, then, in the rhodium-catalyzed industrial process for propylene hydroformylation a high phosphine-to-Rh molar ratio is used. Under these conditions the use of a moderately bulky ligand such as triphenyl phosphine ensures that the catalysis takes place by the topmost cycles in Fig. 5.5, and n-butyraldehyde with high selectivity is produced.

5.2.5   Water Soluble Phosphines and Rhodium Recovery

Rhodium is an expensive metal, and the commercial viability of the rhodium-based hydroformylation process depends crucially on the efficiency of the catalyst recovery process. In the past this has been achieved either by a complicated recycle process or more commonly by energy-requiring distillation. A major advancement in catalyst recovery in recent years has been the introduc-

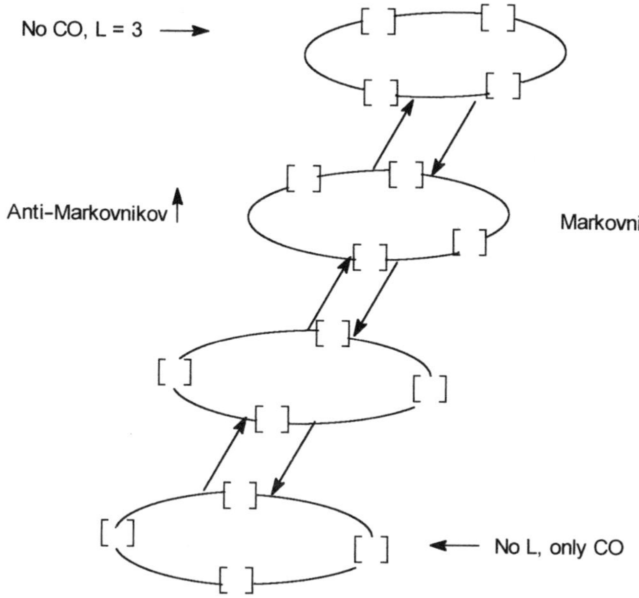

Figure 5.5 Schematic drawing of catalytic cycles coupled by substitutions of L and CO. [ ] are the catalytic intermediates. Figure 5.4 and this figure are basically the same except in the former the cycles were not explicitly shown.

tion of a water-soluble phosphine by Ruhrchemie/Rhone-Poulenc. The phosphorus ligand used is trisulfonated triphenylphosphine, commonly referred to as TPPTS. As shown by 5.7, there is a pH-dependent equilibrium between the water-soluble and the organic-soluble forms of TPPTS.

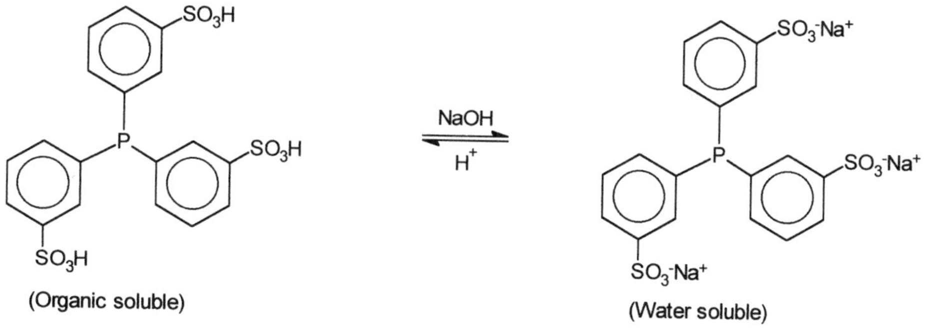

(Organic soluble)   (Water soluble)

(5.7)

The protonated form is extractable with organic solvents between pH 0 and -1, while at higher pH the sodium salt is soluble in water to the extent of 1100 g/liter. The nontoxicity of the ligand (an oral LD_{50} ~5 g/kg) is another feature that makes its large-scale industrial use possible.

Hydroformylation of propylene is carried out in water. The solubility of propylene in water is sufficiently high to give an acceptable rate. The use of a buffer component such as Na_2HPO_4 has been suggested for the control of pH. However, the use of such salts has been shown to have definite influence on the reaction rate as well as product selectivity. The aldehyde product forms an organic layer, which is easily separated by decantation from the catalyst-containing aqueous phase. The high efficiency of the recovery process ensures that Rh losses are in parts per billion ranges. Ruhrchemie/Rhone-Poulenc reports that over a period of 10 years, for the production of 2 million tons of *n*-butyraldehyde, the loss of rhodium has been approximately 2 kg.

The successful commercial use of TPPTS has triggered much research for other water-soluble phosphines. While some of these certainly give highly active and selective catalytic systems, as yet their cost of manufacturing make them commercially unattractive. For example, the *bis*-phosphine 5.14 gives a catalytic system that is about ten times more active than the one with TPPTS and also shows a higher selectivity ($\sim 98\%$) towards *n*-butyraldehyde. Its synthesis, however, is a five-step process and involves expensive reagents.

(5.14)

Recently Union Carbide has reported a separation technique that utilizes monosulfonated triphenyl phosphine (TPPMS). This technique may be useful for the hydroformylation of high molecular weight and less volatile alkenes such as octene, dodecene, and styrene. These alkenes are much less soluble in water than propylene. This means that with the Rh-TPPTS catalytic system in water, unacceptably low rates of hydroformylation are obtained. The use of solubilizing agents such as N-methylpyrrolidone, polyalkylene glycols, etc. makes alkali metal salts of TPPMS soluble in nonpolar organic phase. This probably is due to the formation of reverse micelles, aided by the solubilizing agents. Rhodium complexes of TPPMS can be used to hydroformylate higher olefins in such an organic medium. At the end of the reaction the single phase is separated into a nonpolar and a polar phase by the addition of water or methanol, or by a change in temperature. The catalyst remains in the polar phase, while the product goes into the nonpolar phase.

Another novel technique for separation of the TPPTS-based catalyst is by immobilizing it on a solid support. This catalyst consists of a thin film of an aqueous solution of rhodium–TPPTS complex supported on high-surface-area silica. The substrate alkene is then hydroformylated in an organic solvent such as cyclohexane, which is immiscible with water. The catalytic reaction presumably takes place at the aqueous–organic interface. However, the commercial viability of the Union Carbide discovery and the solid-supported aqueous-phase catalyst remain to be established on an industrial scale.

5.2.6 Catalyst and Ligand Degradation

The lifetime of the rhodium precatalyst depends on the rate at which the metal complex $HRh(CO)(TPPTS)_3$ and the excess ligand TPPTS undergo decomposition. The catalyst lifetime is considerably increased by occasional addition of extra ligand. In general an increase in the reaction temperature and/or CO pressure results in a decrease in the catalyst lifetime.

With 5.1 as the precatalyst, the degradation products in the propylene hydroformylation reaction are diphenylpropyl phosphine ($Ph_2PC_3H_7$), benzene, and less soluble rhodium complexes. These complexes are tentatively formulated as phosphido clusters. Formation of all the products could be rationalized by the hypothetical catalytic cycle shown in Fig. 5.6.

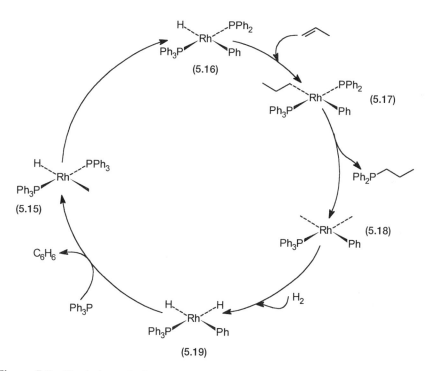

Figure 5.6 Catalytic cycle for the degradation of PPh_3. Only four coordination sites around Rh are shown. In some of the hypothetical intermediates Rh may be five or six coordinated.

The basic assumption is the intramolecular oxidative addition of PPh$_3$ in the coordinatively unsaturated intermediate 5.15 to give the phosphido complex 5.16. This reacts with propylene to give 5.17, which undergoes reductive elimination to give Ph$_2$PC$_3$H$_7$ and the phenyl complex 5.18. The latter undergoes oxidative addition of dihydrogen to give 5.19. This in turn eliminates benzene reductively and regenerates 5.15. The vacant coordination sites of 5.15 and 5.18 are, of course, normally occupied by ligands such as PPh$_3$ or CO. Consequently, the rate of degradation in the presence of excess ligand would be very small.

5.3 COBALT-BASED HYDROFORMYLATION

The catalytic cycle for the cobalt-based hydroformylation is shown in Fig. 5.7. Most cobalt salts under the reaction conditions of hydroformylation are converted into an equilibrium mixture of Co$_2$(CO)$_8$ and HCo(CO)$_4$. The latter undergoes CO dissociation to give 5.20, a catalytically active 16-electron intermediate. Propylene coordination followed by olefin insertion into the metal–hydrogen bond in a Markovnikov or anti-Markovnikov fashion gives the branched or the linear metal alkyl complex 5.24 or 5.22, respectively. These

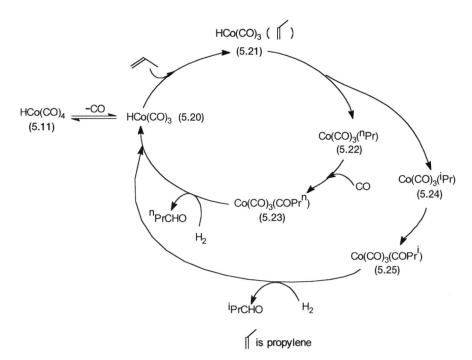

Figure 5.7 Catalytic cycle for the hydroformylation of propylene with cobalt catalyst. The inner and outer cycles show the formation of linear and branched isomers.

complexes can undergo CO insertion into the metal–alkyl bonds to give 5.25 and 5.23, respectively. The product aldehydes are generated from these species by reaction with both dihydrogen and $HCo(CO)_4$. Under the catalytic conditions the latter reaction is insignificant compared to the former.

The cobalt and rhodium catalysts have one important difference between their respective mechanisms. Unlike in the rhodium-catalyzed process, there is no oxidative addition or reductive elimination step in the cobalt-catalyzed hydroformylation reaction. This is reminiscent of the mechanistic difference between rhodium- and cobalt-based carbonylation reactions (see Section 4.2.3). The basic mechanism is well established on the basis of in situ IR spectroscopy, kinetic and theoretical analysis of individual reaction steps, and structural characterization of model complexes.

Both $Co(CO)_4(COPr^n)$ and $Co(CO)_4(COPr^i)$ have been isolated, and their reactions with H_2 as well as $HCo(CO)_4$ have been studied. On the basis of such studies participation by intermediates such as 5.23 and 5.25 in the catalytic cycle is firmly established. The reactions of $HCo(CO)_4$ with $Co(CO)_4(COPr^i)$ and $Co(CO)_4(COPr^n)$ are about 20–30 times faster than the corresponding reactions with dihydrogen at 25°C. However, as already mentioned, at high temperature (>100°C) and pressure (>100 bar), reaction with dihydrogen is the main product-forming step.

For the other catalytic intermediates, there are spectroscopic data and/or strong theoretical arguments in favor of their existence. Thus $Co(CO)_3(COMe)$, an analogue of 5.23 and 5.25 has actually been observed spectroscopically at low temperature by the matrix isolation technique. A similar experimental technique has also established the formation of $Co(CO)_3(Me)$, an analogue of 5.22 and 5.24.

In the Shell process (SHOP) phosphine-modified cobalt-catalyzed hydroformylation is one of the steps in the synthesis of linear alcohols with 12–15 carbon atoms (see Section 7.4.1). Two important characteristics of this reaction should be noted. First, the phosphine-modified precatalyst $HCo(CO)_3(PBu_3^n)$ is less active for hydroformylation than $HCo(CO)_4$ but more active for the subsequent hydrogenation of the aldehyde. In this catalytic system both hydroformylation and hydrogenation of the aldehyde are catalyzed by the same catalytic species. Second, the phosphorus ligand-substituted derivatives are more stable than their carbonyl analogues at higher temperatures and lower pressures (see Table 5.1).

In situ IR spectroscopic studies have been carried out on the $Co_2(CO)_8$ plus PBu_3^n–based catalytic system with ethylene or n-octene as the olefins. Unlike in the case of $Co_2(CO)_8$, where $HCo(CO)_4$ and $Co(CO)_4(COR)$ may be observed, in the presence of PBu_3^n, no $Co(CO)_3(PBu_3^n)(COR)$ or $Co(CO)_3(PBu_3^n))(R)$ types of complexes could be seen. This presumably is because of low concentrations of such intermediates under operating conditions. The mechanism of hydroformylation is assumed to be similar to the one shown in Fig. 5.7. Very little direct mechanistic evidence at a molecular level is available on the aldehyde hydrogenation reaction.

5.4   OTHER HYDROFORMYLATION REACTIONS

Apart from the processes discussed so far, there are other industrial and patented processes where hydroformylation reactions are employed. A few selected ones are summarized in Table 5.2.

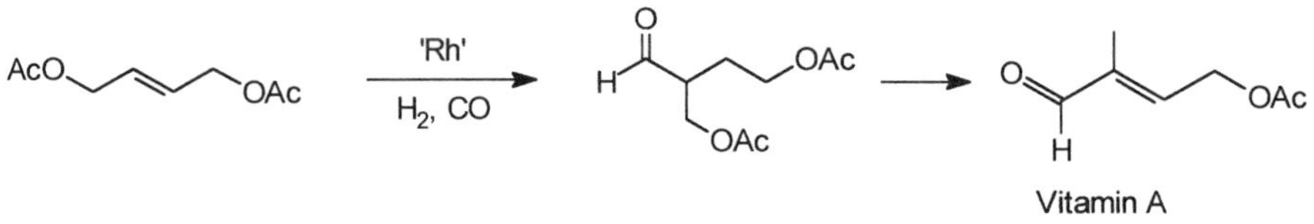

Vitamin A

(5.8)

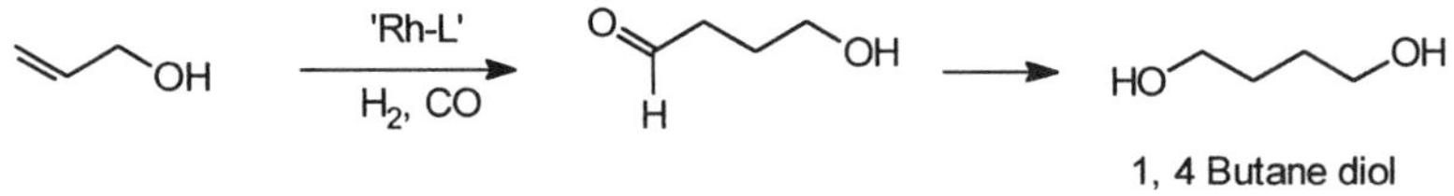

1, 4 Butane diol

(5.9)

3-methyl 1, 5
pentane diol

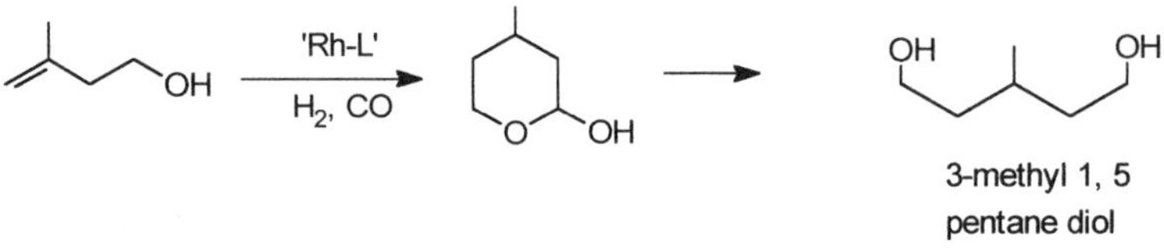

(5.10)

TABLE 5.2   Hydroformylation Reaction Processes

Manufacturer	Product	Process
Mitshubishi Kasei	Isononyl aldehyde for isononyl alcohol, which is used in polyvinyl chloride resin as a plasticizer alcohol	Rhodium catalyst with triphenyl phosphine oxide as a weakly coordinating ligand; catalyst separated from the products by distillation
BASF Hoffman-LaRoche	An intermediate for Vitamin A synthesis	Rhodium catalyst without phosphorus ligand; Reaction 5.8
ARCO	An intermediate for 1,4-butanediol	Rhodium catalyst with chelating phosphorus ligand; hydroformylation of allyl alcohol followed by hydrogenation of the resultant aldehyde; Reaction 5.9
Kuraray	An intermediate for 3-methyl 1,5-pentane diol	$Rh_4(CO)_{12}$ with phosphorus ligand as the precatalyst; hydroformylation of 2-methyl buten-4-ol followed by hydrogenation; Reaction 5.10

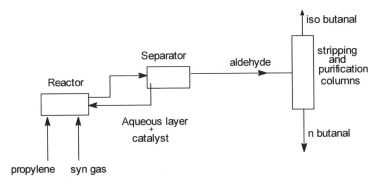

Figure 5.8   Flow sheet of Ruhrchemie/Rhone-Poulenc's process.

5.5   ENGINEERING ASPECTS

In UCC's low-pressure and Mitsubishi Kasei oxo processes the reaction products (isononyl aldehyde, etc.) are separated by distillation from the catalyst phase. As already mentioned, in the Ruhrchemie/Rhone-Poulenc's hydroformylation process, the aqueous phase containing the catalyst is removed after the reaction from the organic phase by decantation. Also in this process, the heat from the oxo reaction is recovered in a falling film evaporator incorporated inside the reactor, which acts a reboiler for the *n*-butanal/isobutanal distillation column.

A simplified flow sheet of the Ruhrchemie/Rhone-Poulenc's process is shown in Fig. 5.8.

PROBLEMS

1. Using hydroformylation and other catalytic or stoichiometric reactions, how could the following transformations be achieved in one or more steps? (a) Ethylene to 2-methylpentanol; (b) butadiene to 1,6-hexanediol; (c) allyl alcohol to butane 1,4-dicarboxylic acid; (d) allyl alcohol to 4-carboxylic butanal.

Ans. (a) Hydroformylation of ethylene, followed by alcohol condensation, followed by hydrogenation; (b) hydroformylation followed by hydrogenation; (c) hydroformylation followed by oxidation; (d) hydroformylation followed by carbonylation.

2. The partial pressures of H_2 and CO in two different catalytic runs of a rhodium-catalyzed hydroformylation reaction were 60 and 20 and 20 and 60 atm, respectively. What effects are expected on the rates?

Ans. If 5.6 holds, then the ratio of rates is 9:1.

3. Suggest an isotope-labeling experiment for characterization of the catalytic intermediate, observed by in situ IR in the hydroformylation of ethylene by cobalt carbonyl.

Ans. $HCo(CO)_4$ is the only species observed. On deuterium labeling a shift in the IR of $\nu_{Co-H(D)}$ by a factor of about 1.44.

4. Treatment of $RhClL_3$ with CH_3CHO gives a complex with IR bands at 1615 and 1920 cm^{-1}, and a high-field twelve-line ^1H NMR signal. This complex on heating gives $RhClL_3$ and CH_3CHO back only if the latter is removed as it is formed. Explain these observations. In what way are they relevant to the catalytic cycle of Fig. 5.1?

Ans. Oxidative addition to give 5.13. The Rh–H and the acyl group absorb at 1920 and 1615 cm^{-1}. δ_{Rh-H} in NMR is coupled to three inequivalent ^{31}P and ^{103}Rh nuclei. Thermodynamically reductive elimination is not a favored reaction. The complex is a model for 5.7 (see D. Milstein, *J. Am. Chem. Soc.*, **104**, 5227–28, 1982).

5. In ethylene hydroformylation with $RhH(CO)(PPh_3)_3$, after several hours of reaction a material containing C, H, and P is isolated that has NMR signals at ^1H (^{31}P) $\sim 7.0\delta$ and a triplet and quartet between 0 and 3δ. What is the material, and how is it formed?

Ans. $Ph_2PC_2H_5$. Degradation of the catalyst by a cycle similar to that of Fig. 5.6.

6. What are the mechanisms of 5.1-catalyzed isomerizations of 2-hexene to 3-hexene, and 1-methyl cyclohexene to 1-methylene cyclohexane? Suggest an experiment.

Ans. For both, insertion followed by β-elimination. For 1-methyl cyclohexene insertion in a Markovnikov manner, while for 2-hexene anti-Markovnikov and Markovnikov are indistinguishable. A *D* label on the methyl of 1-methyl cyclohexene should result in *D* scrambling of the 1,2,6 positions of 1-methylene cyclohexane.

7. Assuming that only the reactions shown in Fig. 5.1 operate for the hydroformylation of propylene to *n*-butyraldehyde with 5.1 as the catalyst, and oxidative addition of dihydrogen is the rate-determining step, what should be the rate expression? What is the implicit assumption?

Ans. Rate expression as in 5.6 with $a = b = 1$ and L instead of CO in the denominator. Assumption: $5.1 + CO + C_3H_6 \rightleftharpoons 5.6 + L$.

8. What qualitative features do you expect in the ^1H NMR spectra of 5.8, 5.9, and 5.10 at room and low temperatures?

Ans. At room temperature all three complexes in rapid equilibrium and one broad signal. At low temperatures: 5.8: ^1H coupled to ^{103}Rh and two equivalent ^{31}P. 5.9: ^{1}H coupled to ^{103}Rh and two inequivalent ^{31}P. 5.10: ^{1}H coupled to ^{103}Rh and one ^{31}P.

9. In reactions 5.8, 5.9, and 5.10 what are the probable steps for the manufacture of the final products from the aldehyde intermediates? In reactions 5.8 and 5.9 what would be the effects of using 5.1 and HRh(CO)$_4$ as the respective precatalysts?

Ans. 5.8: Hydrolysis followed by dehydration, isomerization and acetylation. 5.9, 5.10: Hydrogenation. For 5.8 probably lower rate due to steric hindrance by L but no effect on selectivity. For 5.9 formation of branched isomer (Markovnikov addition).

10. In an industrial hydroformylation reaction with a rhodium catalyst in the presence of excess phosphine and high pressures of CO, what would probably be the minimum number of catalytic cycles and intermediates?

Ans. 4, 24.

11. Many alkenes have low solubility in water. Hydroformylation of such alkenes with Rh-TPPTS catalyst does not give sufficient rate and does not solve the catalyst separation problem. Fluorinated solvents are often immiscible with water and nonfluorinated organic solvents. Suggest a strategy for carrying out hydroformylation of a long-chain alkene in a biphasic system consisting of a fluorinated solvent and another organic solvent.

Ans. Use a phosphine with a fluorinated alkyl group so that a 5.1 type of fluorocarbon-soluble precatalyst is formed in situ (see I. T. Horvath and J. Rabi, *Science* **266**, 72–75, 1994).

12. In Section 5.2.5 immobilization of a thin film of an aqueous solution of Rh-TPPTS complex on high-surface-area silica has been mentioned. This general strategy for some other catalytic reactions sometimes lead to problems due to reaction of water with TPPTS-containing precatalyst. Suggest a solution to this problem.

Ans. Instead of water use some other hydrophilic solvent such as ethylene glycol and an ethylene glycol immiscible organic phase (see K. T. Wan and M. E. Davis, *Nature* **370**, 449–450, 1994).

13. Supercritical carbon dioxide (scCO$_2$) has recently been used as a reaction medium for rhodium-based hydroformylation reactions. In the absence of phosphorus ligands, the rate of hydroformylation of 1-octene is found to be higher in scCO$_2$ than in an organic solvent. Suggest a plausible explanation. What are the potential benefits of scCO$_2$ as a reaction medium?

Ans. Weak solvation leading to easily attained coordinative unsaturation. Unlike organic solvents, scCO$_2$ has no potential environmental hazard (see D. Koch and W. Leitner *J. Am. Chem. Soc.* **120**, 13398–404, 1998).

14. It has been suggested that a dinuclear hydroformylation catalyst where two metal centers mechanistically "cooperate" with each other would be more efficient than a mononuclear one. What type of metal complexes may be used to test this conjecture?

Ans. See M. E. Bronssard et al., *Science* **260**(5115), 1784–88, 1993.

BIBLIOGRAPHY

For all the sections

Books

Most of the books listed under Sections 1.3 and 2.1–2.3.4 contain information on hydroformylation reactions and should be consulted. Specially useful are the book by Parshall and Ittel and Section 2.1.1 of Vol. 1 of *Applied Homogeneous Catalysis with Organometallic Compounds*, ed. by B. Cornils and W. A. Herrmann, VCH, Weinheim, New York, 1996.

Articles

Sections 5.1 to 5.2.2

I. Tkatchenko, in *Comprehensive Organometallic Chemistry*, ed. by G. Wilkinson, F. G. A. Stone, and E. W. Abel, Pergamon Press, Vol. 8, 1982, pp. 101–223.

R. L. Pruett, in *Advances in Organometallic Chemistry*, ed. by F. G. A. Stone and R. West, Academic Press, 1979, Vol. 17, pp. 1–60.

R. L. Pruett, *J. Chem. Edu.* **63**, 196–98, 1986.

T. Onoda, *Chemtech*, **23**(9), 34–37, 1993.

Sections 5.2.3 and 5.2.4

The articles listed above. C. A. Tolman and J. W. Faller, in *Homogeneous Catalysis with Metal Phosphine Complexes*, ed. by L. H. Pignolet, Plenum Press, New York, 1983, pp. 13–109.

For NMR studies on intermediates, see J. M. Brown et al., *J. Chem. Soc. Chem. Commun.*, 721–23 and 723–25, 1982; A. J. Kent et al., *J. Chem. Soc. Chem. Commun.* 728–29, 1985.

For model complexes, see D. Milstein, *J. Am. Chem. Soc.* **104**, 5227–28, 1982.

Section 5.2.5

For TPPS and other water-soluble ligands, see E. G. Kuntz, *Chemtech*, **17**, 570–75, 1987; B. Cornils and E. Wiebus, *ibid.* **25**(1), 33–38, 1995.

W. A. Herrmann and C. W. Kohlpaintner, *Angew. Chem. Int. Ed.* **32**, 1524–44, 1993.

For 5.14 see W. A. Herrmann et al., *J. Mol. Catal A-Chem.* **97**(2), 65–72, 1995.

References given in answers to Problems 11 and 12.

R. V. Chaudhuri et al., *Nature*, **373**, 501–3, 1995.

I. T. Horvath et al., *J. Am. Chem. Soc.* **120**, 3133–43, 1998; A. N. Ajjon and H. Alper, *ibid.* 1466–68.

S. Gladiali et al., *Tetrahedron-Asymmetry* **6**, 1453–74, 1995.

For TPPMS-based Union Carbide work, see *Chem. Eng. News*, April 17, 1995, p. 25.

For more general articles, see D. M. Roundhill, in *Advances in Organometallic Chemistry*, ed. by F. G. A. Stone and R. West, Academic Press, Vol. 38, 1995, pp. 155–88.

B. Driessen-Holscher, in *Advances in Catalysis*, ed. by D. D. Eley, W. O. Hagg, B. Gates, and H. Knozinger, Academic Press, New York, Vol. 42, 1998, pp. 473–506.

Section 5.2.6

A. G. Abatjoglou et al., *Organometallics*, **3**, 923–26, 1984.

Sections 5.3 and 5.4

See articles listed under 5.1 to 5.2.4.

Section 5.5

B. Cornils and E. Wiebus, *Chem. Tech.* (Jan. 1995) 33; Kirk and Othmer *Encyclopedia of Chemical Technology*, 4th edn, John Wiley & Sons, New York, Vol. 16, 1992, pp. 640–51.

CHAPTER 6

POLYMERIZATION

6.1 INTRODUCTION

Polymerization of alkene monomers, with or without functional groups, are very important industrial processes. Until recently the use of homogeneous catalysts was restricted to relatively small-volume production of specialty dimers and oligomers. The manufacture of the two largest-tonnage plastics—polyethylene and polypropylene—has so far been based on heterogeneous catalytic processes.

The importance and relevance of homogeneous catalysis in polymerization reactions have increased tremendously in the past few years for two reasons. First, from about the beginning of the early 1990s a special class of sandwich complexes has been used as homogeneous catalysts. These catalysts, often referred to as metallocene catalysts, can effect the polymerization of a wide variety of alkenes to give polymers of unique properties. Second, the molecular mechanism of polymerization is best understood on the basis of what is known about the chemistry of metal–alkyl, metal–alkene, and other related complexes.

In this chapter we first discuss the salient features of polyethylene and polypropylene manufacture by heterogeneous catalytic processes. We also discuss the structural features of metallocene complexes that are used as homogeneous catalysts and the relationship between the structure of these catalysts and the structures of the resultant polymers.

6.1.1 Polyethylene

Polyethylene was discovered in 1933 by a chance observation in an ICI laboratory. A waxy polymer was found to be formed when ethylene in the presence of benzaldehyde was subjected to high temperature and pressure (170°C, 190 MPa). The resultant polymer had a density of about 0.92 g/cm^3 and was called low-density polyethylene (LDPE). By 1950 three groups—Standard Oil (Indiana), Philips Petroleum, and Karl Ziegler at Max Planck Institute—reported manufacturing processes for polyethylene, at lower temperature and pressure than in the ICI process. The resultant polyethylene of approximate density 0.96 g/cm^3 was called high-density polyethylene (HDPE).

In 1978 Union Carbide reported a special manufacturing process called Unipol that gave linear low-density polyethylene (LLDPE). Linear low-density polyethylene may contain small amounts of butene or octene as co-monomers. The structural differences between HDPE, LDPE, and LLDPE are shown schematically in Fig. 6.1. These structural features determine physical properties such as elasticity, crystallinity, melt-flow index, etc. of the resultant polymers.

6.1.2 Polypropylene

As shown in Fig. 6.2(a) three different types of polypropylene, differing in the orientations of the methyl groups with respect to the polymer backbone, are possible. In isotactic polypropylene the methyl groups are all in one direction, while in atactic polypropylene they are randomly distributed. In syndiotactic polypropylene the orientation of the methyl group alternates in a regular manner.

As shown in Fig. 6.2(b), the relative configurations of the adjacent pseudoasymmetric centers may be designed as meso (m) or racemic (r). Such a designation obviously requires consideration of the stereochemistry of two neighboring units or a dyad. This kind of structural description is useful for determining the degree of iso- and syndiotacticity in a given polymer chain. With a 50-MHz ^{13}C NMR spectrometer all the possible pentad patterns (five monomeric units) with different stereochemistries could be clearly seen. A 150-MHz (^{13}C) machine allows a nonad (nine monomeric units) level analysis. From

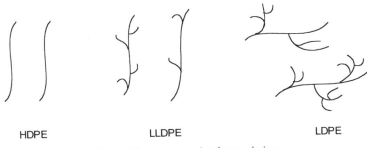

HDPE   LLDPE   LDPE

Curved lines represent polymer chains

Figure 6.1 Schematic representation of high-density (HDPE), linear low-density (LLDPE), and low-density (LDPE) polyethylene molecules.

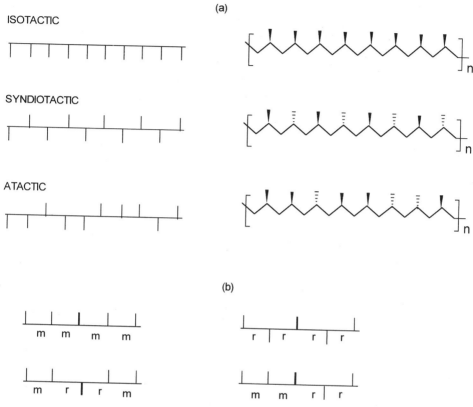

Figure 6.2 Polypropylene of different stereochemistries. In (a) the orientation of the methyl group with respect to the polymer backbone is highlighted. In (b) the stereochemical relationship (meso or racemic) between two adjacent methyl group is shown.

the relative intensities of such signals the degree of tacticity of a given polymer chain can be determined.

Because of its ordered structure, an isotactic polymer has higher melting point and tensile strength than the atactic polymer. The melting points of isotactic polymers are typically between 165 and 171°C, while those of atactic polymers are less than 0°C.

6.2 CATALYSTS FOR POLYETHYLENE

As already mentioned, until about the beginning of 1990 only heterogeneous catalysts have been used for the polymerization of ethylene and propylene. For ethylene polymerization the catalysts used are essentially of three types. These are the Phillips catalyst, the Union Carbide catalyst, and the Ziegler catalyst.

The Phillips catalyst is typically CrO_3 impregnated on silica and calcined at a high temperature (~800°C). The resultant material polymerizes ethylene with an induction time. Another way to prepare the silica-supported chromium catalyst (Union Carbide catalyst) is to react highly reactive chromocene with silica. The reaction between the surface hydroxyl groups of silica and chromocene generates silica-supported organometallic species. Schematically surface anchoring is achieved by reaction 6.1.

$$(6.1)$$

The conventional Ziegler catalyst, named after its inventor, is prepared either by the reaction of $TiCl_4$ with trialkyl aluminum compounds in an organic solvent such as cyclohexane, or from crystalline $TiCl_3$ on an inert support. In the former case the active catalytic species is colloidal in nature and may thus be formally classified as a heterogeneous catalyst. For both chromium- and titanium-based catalysts, choice and pretreatment of the supported catalyst as well as other process parameters determine the molecular weights, distribution of the molecular weights, and the extent of branching of the polymer chains.

6.3 CATALYSTS FOR POLYPROPYLENE

Starting with Natta's discovery that isotactic polypropylene with high crystallinity could be obtained by using a catalyst consisting of $TiCl_3$ and $AlEt_2Cl$, three generations of heterogeneous catalysts for the manufacture of polypropylene have been reported. All these involve $TiCl_3$ and a co-catalyst as two of the essential ingredients. The co-catalyst is an organoaluminum reagent.

$TiCl_3$ generated from the reduction of $TiCl_4$ and $AlEt_3$ exists in four crystalline forms: α, β, γ, and δ. The β form has a chain structure and is brown in color, while the other three have layer structures and are purple in color. The solid-state structures of α and γ may be described as hexagonal and cubic close-packed arrays of chloride ions, respectively. Two-thirds of the octahedral holes of the close-packed arrays are filled by Ti^{3+} ions. The δ form is more disordered than both α and γ.

The polymorphs α, γ and δ give polypropylene with high isotacticity, but β does not. The organoaluminum compound is a vital ingredient, since it generates the titanium alkyl precursor from $TiCl_3$. Formation of the latter from $TiCl_4$ takes place by the following reactions:

$$6TiCl_4 + 2AlEt_3 \rightarrow 6TiCl_3 + 2AlCl_3 + 3C_2H_6 + 3C_2H_4 \qquad (6.2)$$

$$6TiCl_4 + 3AlEt_2Cl \rightarrow 6TiCl_3 + 3AlCl_3 + 3C_2H_6 + 3C_2H_4 \qquad (6.3)$$

The monoethyl reagent, $AlEtCl_2$, is not active and in fact is a poison probably because it blocks the coordination site (see Section 6.4). However, in the absence of promoters, often referred to as the third component, $AlEt_2Cl$ is consumed and $AlEtCl_2$ is formed according to reaction 6.4. Promoters such as ethers, esters, etc. are added to destroy $AlEtCl_2$ and to prevent its formation by reactions such as 6.5.

$$AlEt_2Cl + AlCl_3 \rightarrow 2AlEtCl_2 \qquad (6.4)$$

$$2AlEtCl_2 + R_2O \rightarrow AlCl_3 \cdot OR_2 + Et_2AlCl \qquad (6.5)$$

The third-generation catalyst currently used worldwide consists of $TiCl_4$ supported on $MgCl_2$. It is used in combination with an organoaluminum reagent and a suitable third component. Note that $MgCl_2$ also has a layered structure and the ionic radii of Mg^{2+} and Ti^{3+} ions are very similar. This structural compatibility and closeness are considered to be responsible for the high activity of the third-generation catalysts. Productivity as high as 30,000 g of polypropylene per gram of catalyst per hour could be achieved with an isotacticity index of 96–99%.

Synthesis of the third-generation catalyst is an intricate process. Two crucial steps are: (1) Precipitation of the supported catalyst from a solution of Mg^{2+} ion in organic solvent by the addition of $TiCl_4$. (2) Catalyst activation by heat treatment with $TiCl_4$ and phthalate esters (third component). The precursors for the support could be magnesium alkoxides, carboxylates, sulfite, or sulfinates. They all give particles of different but well-defined morphologies.

6.4 CATALYTIC CYCLE FOR ALKENE POLYMERIZATION

Two major mechanisms have been proposed for alkene polymerization. These are the Cossee–Arlman mechanism and the Green–Rooney mechanism. A modified version of the latter has also been considered to explain the behavior of homogeneous, metallocene catalysts. The original Cossee–Arlman mechanism was proposed for the $TiCl_3$ based heterogeneous catalyst. In the following sections we discuss these different mechanisms in some detail. In the following discussion in accordance with the results obtained from the metallocene systems, the oxidation states of the *active surface sites* are assumed to be 4+.

6.4.1 Cossee–Arlman Mechanism

The Cossee–Arlman mechanism proposes direct insertion of alkene into the metal–alkyl bond (see Section 2.3.2) without the formation of any intermediate. In the solid catalyst anion vacancies at the crystal edges are formed by simple cleavage of the bulk. Some of the Ti^{4+} ions located at these surfaces are thus coordinatively unsaturated (i.e., five rather than six coordinate). Moreover, steric constraints arising due to the presence of the surface anions make the coordination of an alkene such as propylene stereospecific. This gives rise to isotacticity in the product polypropylene. For the time being we defer any discussion on the tacticity aspect and confine our attention to the basic mechanism of polymer growth. A catalytic cycle for ethylene polymerization according to the Cossee–Arlman mechanism is shown in Fig. 6.3.

Coordinatively unsaturated 6.1 represents the surface Ti^{4+} responsible for catalyzing the polymerization reaction. By reaction with the Et_3Al (or Et_2AlCl), a Ti–Et bond, as shown for 6.2, is formed. Ethylene coordination at the vacant

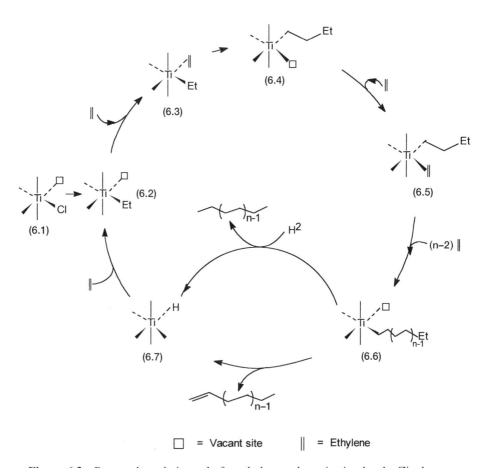

□ = Vacant site   ‖ = Ethylene

Figure 6.3 Proposed catalytic cycle for ethylene polymerization by the Ziegler catalyst. From 6.5 to 6.6 $n - 2$ ethylene molecules undergo insertion. Note the alteration of coordination sites between the polymer chain and coordinated ethylene, that is, migratory insertion, is assumed.

site of 6.2 produces 6.3. The subsequent step, conversion of 6.3 to 6.4 is crucial. It is here that direct insertion of ethylene into the metal–alkyl bond takes place. As shown by 6.5 the vacant coordination site thus created is occupied again by another molecule of ethylene. Conversion of 6.5 to 6.6 is basically a repeat of the two preceding steps $n - 2$ times. The polymer chain grows by continuous coordination followed by insertion or, more accurately, migratory insertion. Conversion of 6.6 to 6.7 may occur by two reactions. If hydrogen is deliberately added, then cleavage of the metal–alkyl bond, and formation of a metal–hydrogen bond, frees the saturated polymer. Alternatively, 6.6 undergoes β-hydride elimination, leading to the formation of a polymer chain with an unsaturated end group and the metal–hydrogen bond.

As far as the fate of 6.6 is concerned, there is a third possibility. The polymer chain may remain attached to the metal atom; that is, the metal–alkyl bond remains intact. The product of the overall reaction in this case is 6.6. Polymerization of this type is often called *living polymerization*. Note that in this case there is no polymer chain termination step, and the catalytic cycle is not completed. In other words, although a single Ti^{4+} ion may be responsible for the polymerization of thousands of ethylene molecules, in the strictest sense of the term it is not a true catalytic reaction. In the event 6.6 is converted to 6.7 either by reaction with H_2 or by β-elimination, 6.7 further reacts with ethylene. Insertion of ethylene into the Ti–H bond regenerates 6.2 and completes the catalytic cycle.

6.4.2 Mechanism of Alkene Insertion

In the Cossee–Arlman mechanism insertion is considered to be direct. The *transition state* of 6.5 to 6.6 by the Cossee–Arlman mechanism is therefore as designated by 6.8. In 6.8, for clarity the Cl^- ligands are not shown and Ⓟ represents the growing polymer chain.

(6.8)

As we saw in Section 2.2.2, an interaction between metal and the α-hydrogen of an alkyl group is called an *agostic interaction*. In an extreme situation, where the interaction between the metal atom and the α-hydrogen leads to formal cleavage of the carbon–hydrogen bond, a mechanism involving a metal–carbene *intermediate* may be invoked. This proposal is known as the Green and Rooney mechanism, and two of the proposed catalytic intermediates are shown by 6.9 and 6.10.

$$\left[\begin{array}{c} \overset{H}{\underset{|}{\gt}}\hspace{-2mm}Ti\hspace{-1mm}=\hspace{-3mm}\overset{H}{\underset{\diagdown}{\cancel{\diagup}}}\hspace{-2mm}(P) \end{array} \right]$$

(6.9)

$$\left[\begin{array}{c} \overset{H}{\underset{|}{\gt}}\hspace{-2mm}Ti\hspace{-2mm}\overset{H\;(P)}{\diagup\!\!\diagdown} \end{array} \right]$$

(6.10)

The heterogeneous character of the conventional Ziegler–Natta catalyst makes studies directed towards mechanistic and structural elucidation at a molecular level extremely difficult. Experimental evidence is therefore sought from homogeneous metallocene and other related catalysts (see Section 6.5). Such evidence does not support the Green–Rooney mechanism.

A mechanism as outlined by Fig. 6.4 is, however, possible. Here agostic interactions in the transition state, 6.12, and the subsequent catalytic intermediate, 6.13, are present. Indeed, for certain homogeneous catalytic systems there is good experimental evidence for agostic interactions. It is important to note that the difference between 6.8 and 6.12, the two transition states, is only in the presence or otherwise of agostic interaction. Also in 6.13 the agostic interaction is with an H atom on the γ-carbon. As we will see, agostic interactions are indeed responsible for the high stereospecificity of propylene polymeriza-

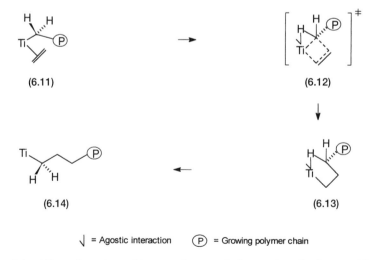

$\sqrt{}$ = Agostic interaction   (P) = Growing polymer chain

Figure 6.4 Alkene insertion with α- and γ-agostic interactions in the transition state 6.12 and intermediate 6.13. Such interactions are important for stereoregular polymerization by metallocene catalysts.

tion with metallocene catalysts. For ethylene polymerization agostic interactions do not have any practical consequences.

6.4.3   Mechanistic Evidence

The mechanism for polymerization of propylene with heterogeneous catalysts is very similar to that of ethylene. Studies with a homogeneous catalyst of a lanthanide element provided early mechanistic evidence. The complex used in these studies was 6.15. In 6.15 lutetium is in a 3+ oxidation state and has the electronic configuration of $4f^{14}$. In other words Lu^{3+} has a full f shell and 6.15 is a diamagnetic complex.

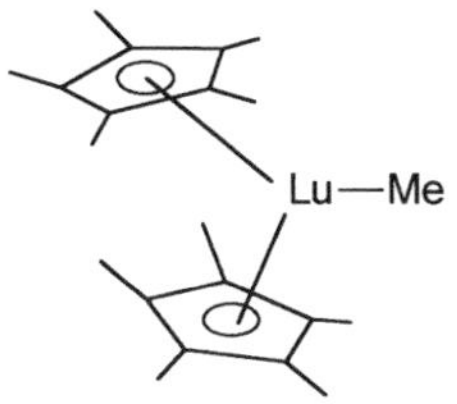

(6.15)

Apart from providing direct spectroscopic evidence for the various catalytic steps, experiments with 6.15 and other lanthanide cyclopentadiene complexes also established that paramagnetic metal ions do not have any special effect on the oligomerization reaction. The mechanism proposed on the basis of these studies is as shown in Fig. 6.5.

The conversions of 6.16 to 6.17 and 6.19 to 6.20 are the chain-propagation steps. The conversions of both 6.17 and 6.20 to 6.18 by β-hydride elimination are the chain-termination steps. The last two reactions are also the ones where the product polypropylene is formed. Conversion of 6.16 to 6.18 without the intermediacy of 6.17 also involves β-hydride elimination, and the product formed is isobutylene.

6.5   METALLOCENE CATALYSTS

The discovery of homogeneous metallocene catalysts in the 1980s was a very important milestone in polymer technology. With these catalysts the plastic industry is poised to move into an era of an entirely new range of polymeric materials with several specific advantages. From the initial discovery in the 1980s, close to five billion dollars is estimated to have been invested by several large chemical companies in research and development. In a relatively short time, this has resulted in approximately fifteen hundred patent applications! Close to 0.5 million tons of metallocene-catalyzed polypropylene is expected to be manufactured by the year 2003.

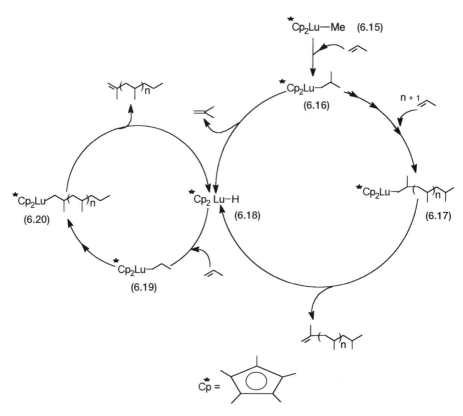

Figure 6.5 Mechanism and catalytic cycle for propylene polymerization with a model metallocene catalyst. Conversion of 6.16 to 6.17 and 6.19 to 6.20 involve insertion of (*n* + 1) propylene molecules.

Sandwich or metallocene complexes have been known for almost 50 years. However, only in the 1970s was it discovered that when trimethylaluminum was used as a co-catalyst, small amounts of water greatly increased the polymerization activity of a metallocene catalyst. As we shall see in the following sections, this apparently simple observation has an interesting mechanistic explanation. We first discuss the general structures of the metallocene catalysts and the co-catalysts.

6.5.1 Structures of Metallocene Catalysts and the Co-Catalysts

A general structure that describes all metallocene catalysts cannot be given since a variety of bicyclopentadineyl and monocyclopentadineyl complexes have been used as catalysts. A general structure for bicyclopentadienyl complexes used as catalysts is shown by 6.21.

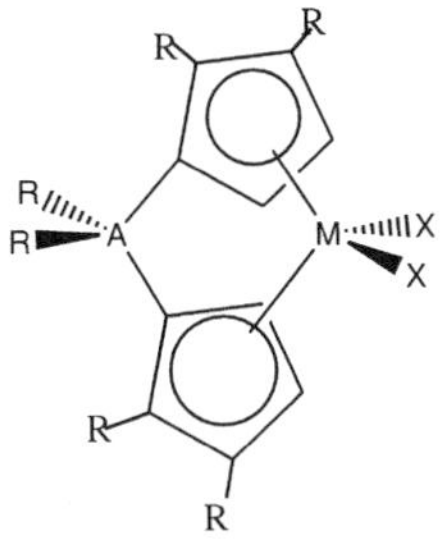

(6.21)

The following points need to be made. First, the five-membered ring may also be a part of an indenyl ring structure, and the two five-membered rings may be indenyl ligands or one may be an indenyl while the other is a cyclopentadineyl ring. Second, the metal (M) is titanium or zirconium in the oxidation state of four. Third, A is an optional bridging atom, generally a carbon or silicon atom with R=CH$_3$, H, alkyl, or other hydrocarbon groups. The Rs on A and on the five-membered rings need not necessarily be the same. Finally, X is usually Cl or an alkyl group. Two typical examples of this type of catalyst are shown by structures 6.22 and 6.23. As we shall see later, the structural difference in terms of symmetry between 6.22 and 6.23 has an important bearing on the tacticity of polypropylene produced by these catalysts.

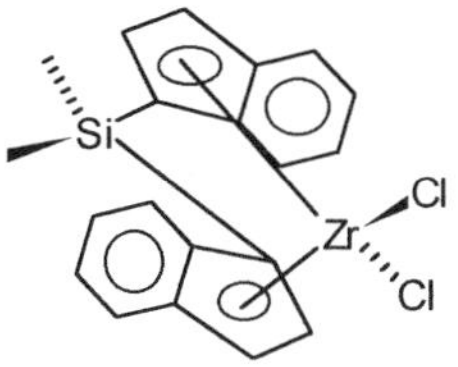

(6.22)

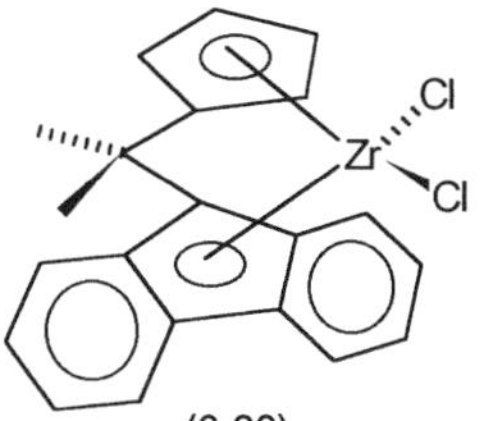

(6.23)

Monocyclopentadienyl complexes have also been used as active polymerization catalysts. Indeed, titanium monocyclopentadienyl metallocenes—the so-called "constrained geometry catalysts"—have been focal points of Dow's activity in this area. A typical example of such a catalyst is shown by the structure 6.24.

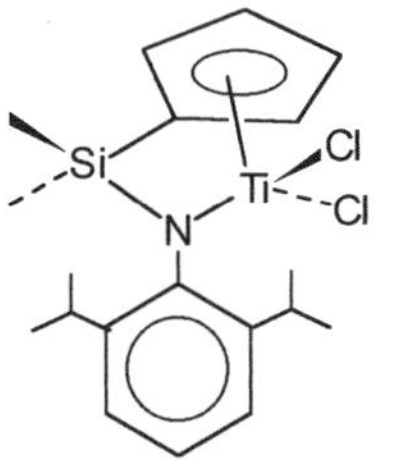

(6.24)

As mentioned earlier, the soluble metallocene catalysts exhibit high polymerization activities only in combination with water and Me$_3$Al. The reaction between Me_3Al and water generates methyl aluminoxanes (MAO), which are the actual active co-catalysts. The function of these co-catalysts is the generation of a coordinatively unsaturated catalytic intermediate. This is shown schematically in Fig. 6.6.

Here MAO first generates the dimethyl complex 6.26 from 6.25. This reaction, of course, can also be brought about by Me_3Al. It is the subsequent reaction (i.e., the conversion of 6.26 to 6.27 that is of crucial importance. The high Lewis acidity of the aluminum centers in MAO enables it to abstract a CH_3 group from 6.26 and sequesters it in the anion, $[CH_3\text{-}MAO]^-$. Although 6.27 is shown as ionically dissociated species, probably the anion, $[CH_3\text{-}MAO]^-$, weakly coordinates to the zirconium atom. It is this coordinatively unsaturated species, 6.27, that promotes the alkene coordination and insertion that are necessary for polymerization activity.

The general formula of MAO is given by $Me_2Al\text{-}O\text{-}[AlMe]_n\text{-}OAlMe_2$, where n is between 5 and 20. Controlled hydrolysis of $AlMe_3$ produces a mixture of such oligomers. The solution structures of these oligomers have been investigated by a variety of spectroscopic and other techniques, including ^{27}Al NMR. From such studies intramolecular μ_3-O bridges between three aluminum atoms 6.28, and μ_2-CH_3 bridges between two aluminum atoms 6.29, have been identified.

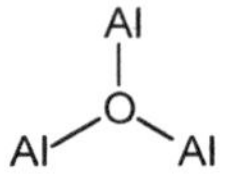

(6.28)

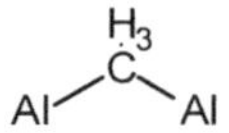

(6.29)

Also present are both four- and three-coordinate Al centers, the latter probably giving rise to high Lewis acidity. Structures of some aluminoxane clusters have also been determined in the solid state by single-crystal X-ray studies. One such structure, that of $[Al_7O_6Me_{16}]$ anion, is shown by 6.30. Another structure, 6.31, has been proposed on the basis of mass spectroscopic data. All the evidences taken together point to open-cage rather than dense mineral structures for the aluminoxanes in MAO solutions.

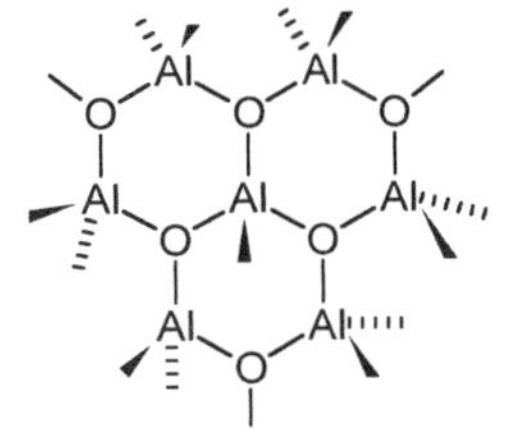

(6.30)

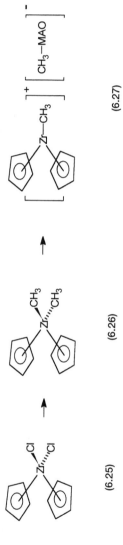

(6.25)　　　　　　(6.26)　　　　　　(6.27)

Figure 6.6 Production of active catalyst 6.27 from the precatalyst 6.25 by treatment with methyl alumino oxane. 6.26 does not have free coordination site and cannot act as a catalyst.

(6.31)

6.5.2 Special Features and Advantages of Metallocene Catalysts

The mechanical, thermal, optical, and other properties of a polymer depend on the structure of the monomer units. Where a copolymer is used, it depends additionally on the relative amounts and distribution of the two monomeric building blocks. Metallocene catalysts have four main advantages over the conventional polymerization catalysts. They can polymerize a very wide variety of vinyl monomers irrespective of their molecular weights and steric features. They also can polymerize mixtures of monomers to give polymers of unique properties.

The second advantage of a metallocene catalyst is the advantage of the single site. Microcrystalline Ziegler–Natta catalyst, with nonidentical coordination environments around the catalytically active metal centers, give polymers with broad molecular-weight distributions. Metallocene catalysts, on the other hand, produce extremely uniform homopolymers and co-polymers. This is because every molecule of the metallocene catalyst has an identical coordination environment and therefore produces identical or nearly identical polymer chains. If a broad molecular-weight distribution is desired, that can be achieved with high precision by using more than one catalyst and/or cascade reactor configurations. In the latter case different reactors are operated under different conditions to produce polymers of different molecular weights.

The third advantage associated with metallocene catalysts is that the predominant mechanism for chain termination is by β-hydride elimination. This produces a vinyl double bond at the end of each polymer chain. Further functionalization of the vinyl group by graft polymerization with maleic anhydride and other functional monomers is far more effective than is typical for polyolefins obtained by conventional catalysts.

Finally, the most important feature of the metallocene catalysts is their ability to produce highly stereoregular polymers. The molecular geometry of the metallocene molecule directly controls the stereoregularity of the resultant polymer. By the correct choice of ligand environment it is possible to generate highly stereoregular polypropylene or polystyrene. Indeed, syndiotactic polystyrene and other aromatic polymers are a new class of materials made available only by metallocene catalysts.

6.5.3   Mechanism of Polymerization and Stereocontrol by Metallocene Catalysts

The basic mechanism of metallocene-based polymerization involves a catalytic cycle very similar to that of Fig. 6.5. The precatalysts 6.22 and 6.23, in combination with MAO, produce polypropylene of high isotacticity and syndiotacticity, respectively. As shown in Fig. 6.7, 6.22 has C_2 symmetry and is chiral, while the symmetry of 6.23 is C_s and is therefore achiral. Two points need to be noted before we discuss the mechanism of stereospecific insertion of propylene. First, propylene is a planar molecule that has two potentially nonequivalent, prochiral faces (see Section 9.3.1). Second, the symmetry around the metal atom determines whether or not coordinations by the two faces of propylene are equivalent.

The mechanism of stereocontrol has been investigated by a variety of techniques. These include ^{13}C NMR, molecular mechanics, ab initio calculations, and kinetic studies with deuterium-labeled alkenes. From these studies the following conclusions can be drawn. First, the process of stereocontrol begins only after the growing polymer chain acquires at least two carbon atoms. For a chiral catalyst like 6.22, every single molecule of propylene coordinates to the metal atom only through one specific prochiral face, and not the other one. Second, the metal alkyl chain favors an orientation where its $C(\alpha)–C(\beta)$ segment is in the most open sector of the metallocene catalyst. In this orientation the substituent on the alkene (i.e., the methyl group of the coordinated propylene molecule) is *trans* to the β-carbon atom of the metal bound alkyl chain. In other words, it is far away from the alkyl chain as possible. This is shown schematically in Fig. 6.8. Finally, as mentioned earlier (see Section 6.4.2), the transition state is stabilized by an α-agostic interaction. Note that in the absence of agostic interaction rotation around the Zr—C bond would cause the *trans* orientation between the methyl substituent of propylene and (P) of the growing polymer chain to be lost.

Two examples clearly illustrate the relationship between molecular structures of the metallocene catalysts on the one hand, and the tacticity of the resultant polymers on the other. As shown in Fig. 6.9, complexes 6.32, 6.33, and 6.34 have very similar structures. In 6.33 and 6.34 the cyclopentadiene ring of 6.32 has been substituted with a methyl and a *t*-butyl group, respectively. The effect of this substitution on the tacticity of the polypropylene is remarkable. As already mentioned, 6.32, which has C_s symmetry, gives a syndiotactic polymer. In 6.33 the symmetry is lost and the chirality of the catalyst is reflected in the hemi-isotacticity of the polymer, where every alternate methyl has a random orientation. In other words, the insertion of every alternate propylene molecule is stereospecific and has an isotactic relationship. In 6.34 the more bulky *t*-butyl group ensures that every propylene molecule inserts in a stereospecific manner and the resultant polymer is fully isotactic.

The second example uses a precatalyst that has a temperature-dependent equilibrium between two structures—one chiral and the other achiral. This is

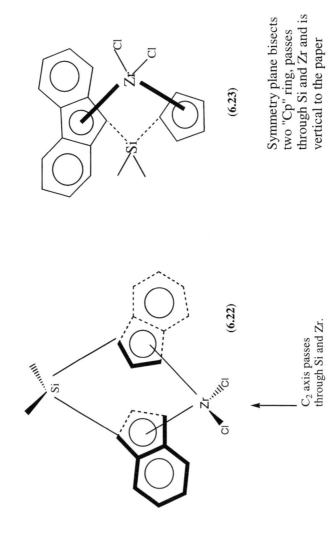

(6.22)

C$_2$ axis passes
through Si and Zr.

(6.23)

Symmetry plane bisects
two "Cp" ring, passes
through Si and Zr and is
vertical to the paper

Figure 6.7 Schematic presentation of metallocene catalysts of C_2 and C_s symmetry. The dark and broken lines represent above and below the paper.

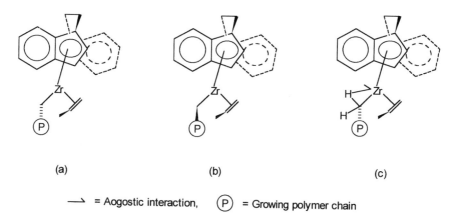

| (a) | (b) | (c) |

⟶ = Aogostic interaction,   Ⓟ = Growing polymer chain

Figure 6.8 Stereospecific propylene insertion in a metallocene catalyst of the type 6.22. For clarity the coordination of Zr to the second indene ring (broken line) is not shown. (a) Preferred orientation of the growing polymer chain. Note the *trans* orientation of the methyl group (above the plane of paper) and Ⓟ (below the plane of paper). (b) Rotation along the Zr—C bond may make Ⓟ and CH_3 *cis* to each other. (c) Agostic interaction that prevents rotation around the Zr—C bond and keeps Ⓟ and CH_3 away from each other.

shown in Fig. 6.10. The use of this precatalyst in combination with MAO produces polypropylene with blocks of "isotactic" and "atactic" fragments. The relative amount of such blocks is determined by the relative amounts of 6.35 and 6.36 that are present at a given temperature.

Note that at lower temperatures the amount of 6.36 present would be less compared with 6.35. This is because of the "eclipsed" structure of 6.36, which has higher nonbonded interactions than that of 6.35. The net effect of this is that at lower temperatures more isotactic blocks are formed. Thus the resultant polymer shows a higher degree of isotacticity than that of polymers formed at ambient or higher temperatures.

Finally, as already mentioned, metallocene catalysts can polymerize a variety of olefins. In certain cases the structural features of the monomer lead to the formation of novel polymers. Two such examples are shown by reactions 6.6 and 6.7. It is clear that the polymerization processes involve considerable rearrangements of the bonds. Reactions 6.8 and 6.9 show the formal mechanisms of such rearrangements for 1,5-hexadiene and methylenecyclobutane, respectively.

$$\text{(6.6)}$$

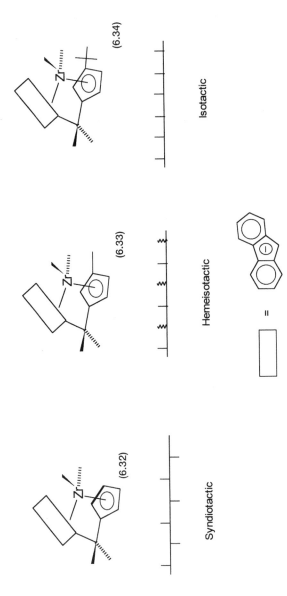

Figure 6.9 Effect of symmetry and substituents on the stereochemistry of the resultant polypropylene. 6.32 has C_s symmetry. 6.33 is chiral, but the effect of Me is moderate. 6.34 is also chiral, and the effect of bulky Bu^t is more marked.

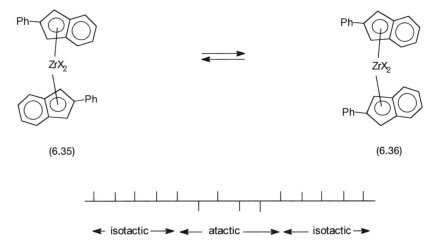

Ph—   ZrX$_2$ —Ph $\rightleftharpoons$ Ph—   ZrX$_2$ Ph—

(6.35)   (6.36)

←— isotactic —→ ←— atactic —→ ←— isotactic —→

Figure 6.10   Temperature-dependent equilibrium between chiral and achiral isomers of a metallocene catalyst and the resultant polypropylene with isotactic and atactic polymer-blocks.

(6.7)

M   M   P   (6.8)

P = Polymer

M   M   (6.9)

6.6   CHROMOCENE AND HETEROGENEOUS CATALYSTS

In Section 6.2, it was mentioned that one of the commercial catalysts for poly-ethylene (HDPE) is made by the reaction of chromocene with silica according to reaction 6.1. The oxidation state of the catalytically actives sites has been the subject matter of some debate. Recent experiments with well defined homogeneous cyclopentadiene chromium complexes, 6.37–6.39, strongly suggest that the oxidation state of chromium in this catalyst is 3+ rather than 2+.

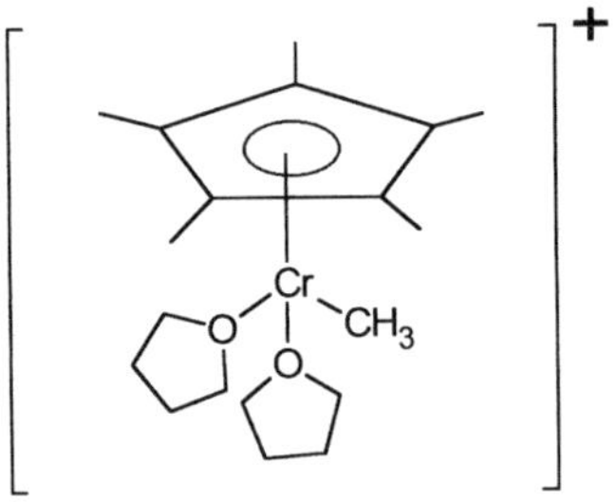

(6.37)

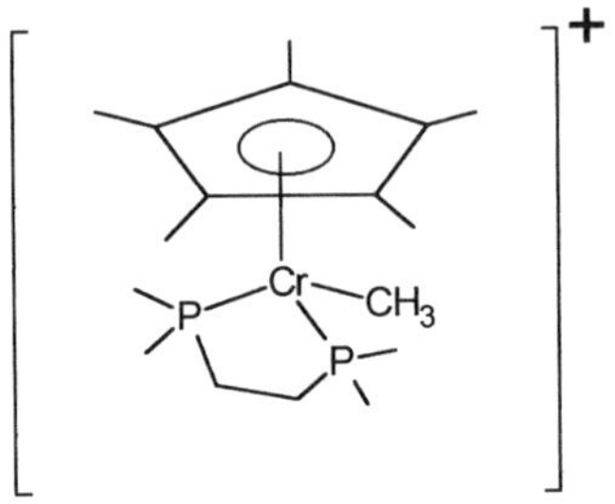

(6.38)

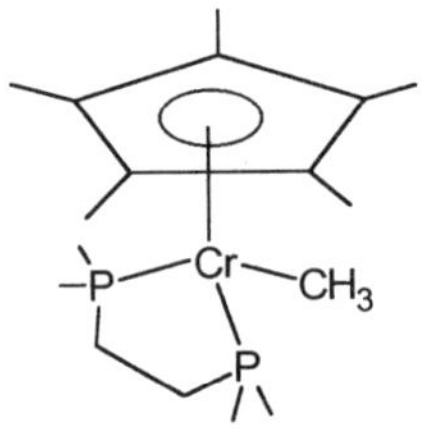

(6.39)

Complex 6.37 has been shown to be an active catalyst for the manufacture of HDPE. Note that 6.37 is a 15-electron complex with chromium in the formal oxidation state of 3+. The mechanism of polymerization involves generation of coordinative unsaturation through the dissociation of a THF molecule from 6.37. The evidence for an oxidation state of 3+ in the commercial catalyst comes from the fact that complex 6.38 is active for polymerization. However, complex 6.39, identical to 6.38 in every respect except the oxidation state of the metal ion, is inactive. Note that the oxidation state of chromium in 6.39 is 2+.

6.7   POLYMERS OF OTHER ALKENES

So far we have restricted our discussions mainly to polyethylene and polypro-pylene—the two most important and largest-capacity polymers commercially produced. There are a number of other homo- and co-polymers derived from a variety of alkenes, which are used as polymers for special purposes. The commercial routes for most of these polymers involve heterogeneous catalysts. The mechanisms at the molecular level are likely to be very similar to the ones discussed so far. A few of these speciality polymers are now being manufac-tured by truly homogeneous, metallocene catalysts, and many more are ex-pected to be made in the near future. In Table 6.1 a summary of the properties, uses, and the catalysts required for these speciality polymers is given.

6.8   ENGINEERING ASPECTS

The Unipol process employs a fluidized bed reactor (see Section 3.1.2) for the preparation of polyethylene and polypropylene. A gas–liquid fluid solid reactor, where both liquid and gas fluidize the solids, is used for Ziegler–Natta cata-lyzed ethylene polymerization. Hoechst, Mitsui, Montedison, Solvay et Cie, and a number of other producers use a Ziegler-type catalyst for the manufacture of LLDPE by slurry polymerization in hexane solvent (Fig. 6.11). The system consists of a series of continuous stirred tank reactors to achieve the desired residence time. 1-Butene is used a comonomer, and hydrogen is used for con-trolling molecular weight. The polymer beads are separated from the liquid by centrifugation followed by steam stripping.

DuPont and Dow use solution polymerization technology to produce LLDPE resins. The process is based on continuous polymerization of ethylene with 1-octene in cyclohexane at about 250°C and 1200 psi. The catalyst is again Ziegler type. Residence time is of the order of several minutes. The catalyst is deactivated by treatment with an alcohol or complexing agent such as acetyl-acetone, and adsorbed on a silaceous adsorbent before stripping the solvent. The Stamicarbon (Dutch State mines) process is similar to the DuPont process, and it uses a short-residence-time solution process for HDPE production.

Ziegler catalysts are also used for the manufacture of HDPE at low (40–150 psi) or medium (290–580 psi) pressures. The polymerization is carried out in a stirred reactor at about 80–90°C and 145 psi, or in a loop reactor at about 450 psi and temperatures between 65 and 110°C. Since the catalyst is very efficient, its concentration is kept very low, obviating the need for catalyst deactivation and removal steps. Conversions are very high. The solvent is re-moved by centrifugation, while the polymer is dried in a fluidized bed drier.

Various procedures followed for the manufacture of polypropylene are sum-marized in Table 6.2.

TABLE 6.1 Specialty Polymers

Polymer	Catalyst	Properties and uses
Ethylene, propylene, diene, (EPDM) rubber	1. $VOCl_3/AIR_2Cl$ 2. $ZrCp_2Cl_2/MAO$	Elastic polymer (elastomer); can be cured by conventional rubber technology; the diene provides two double bonds of differing reactivity; the less reactive one is used in curing
cis-1,4-Polybutadiene	Ziegler—Natta-type catalysts	Synthetic rubber; of four possible configurations the cis-1,4-conformer is 90–93%.
Poly-4-methyl-1-pentene (one of the dimers of propylene)	Ziegler–Natta type to give stereoregular polymer	Highly transparent and very low-density polymer; used in making medical instruments and laboratory wares
Poly(1-decene); poly(1-dodecene)	Ziegler–Natta type	Very high-molecular-weight polymers used in parts per million levels to improve flow properties of oil in oil pipelines

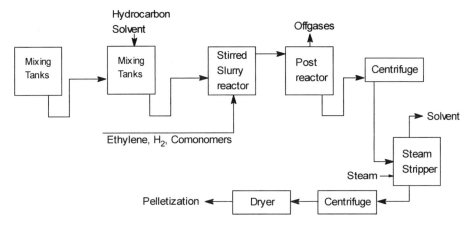

Figure 6.11   Schematic diagram of a slurry polymerization process.

TABLE 6.2   Polypropylene Manufacture

Manufacturer	Reactor (conditions)	Comments
Hercules, Eastman, and Shell	Solvent slurry reactor; reaction conditions: about 50–90°C, 75–225 psi	Unreacted monomer recovered by flash distillation, catalyst by extraction, and polymer particles by centrifugation
Exxon and Phillips	Tabular reactor with liquid monomer	Expensive process but gives polymer of high purity
Montedison, Mitsui	Monomer slurry reactor	Relatively simple process with high product yield

PROBLEMS

1. In Fig. 6.5 label each step of the cycles as initiation, propagation, or elimination.

Ans. Initiation: 6.16 to 6.18; 6.15 to 6.16; propagation: 6.16 to 6.17, 6.19 to 6.20; termination: 6.17 to 6.18, 6.20 to 6.18.

2. In Question 1, in the absence of termination steps, what would be the products?

Ans. Living polymerization, polymers/complex of the type 6.17, 6.20.

3. What are the symmetries of 6.24 and 6.36?

Ans. C_1 and C_{2v}.

4. In the m, r notation how many different signals are expected for a given "pentad" and "nonad" that have all the possible configurations? How are their isotacticities determined?

Ans. $^5C_2 = 10$, $^9C_2 = 36$. For the pentad, ^{13}C NMR intensity for mmmmm/ combined ^{13}C NMR intensities of all the others. Similarly for the nonad.

5. In 6.27 the CH_3 group in $[CH_3\text{-MAO}]^-$ is formally treated as an anion. In 6.30, an anion with a single negative charge, how many "anionic" CH_3 groups are present?

Ans. 13, bonded to 7 Al atoms.

6. Of the two types of surface-supported chromium in reaction 6.1, which one is expected to be catalytically active?

Ans. The one bonded to single oxygen as it has coordinative unsaturation.

7. Treatment of 6.38 with Na–Hg followed by reaction with ethylene gives a product with a molecular ion peak in the mass spectrum at 42. What could be concluded from this?

Ans. Reduction by Na–Hg gives 6.39. Reaction of the latter with ethylene gives propylene because of facile β-elimination rather than chain propagation.

8. A transition metal complex catalyzes the trimerization of ethylene to give 1-hexene. Draw a catalytic cycle like the one shown in Fig. 6.3 for this reaction.

9. Draw the expected structure of the polymer chain obtained from the polymerization of cis-$CD_3CD{=}CHD$ by the Ziegler–Natta catalyst. What could be an alternative configuration? What mechanistic information is gained from such an isotope-labeling experiment?

Ans. CD_3 and H cis to each other in the polymer chain; that is, they have "erythro" configuration with respect to each other. With Natta catalyst a high isotacticity index is expected. An alternative configuration would be "threo" (CD_3 and H trans to each other). The observed "erythro" configuration means alkene insertion is cis.

10. In Fig. 6.5, conversions of 6.18 to 6.19, 6.19 to 6.20, and 6.16 to 6.17 assume a certain regioselectivity. What is the assumption and what would be the consequences if the assumption was not valid?

Ans. All anti-Markovnikov insertions. Instead of every third (alternate) carbon having a methyl substituent, adjacent carbons would also have methyl substitution in a random manner.

11. The polymer obtained from 1,5-hexadiene using 6.22 and MAO has measurable optical rotation. Why?

Ans. Refer to reaction 6.6. 6.22 is expected to give an isotactic polymer. On steric grounds trans-isotactic configurations of 1,5 and 3,4 bonds are expected. This configuration, as opposed to *cis*-isotactic, is chiral (see G. W. Coates and R. M. Waymouth, *J. Am. Chem. Soc.* **115**, 91–98, 1993).

12. The ratio of the molecular weights of polymers obtained from (*E*)-1-propene-1-d_1 and (*Z*)-1-propene-1-d_1 is found to be about 1.3. Can this be explained in terms of α-agostic interactions and kinetic isotope effect?

Ans. The ratio of molecular weights is a measure of relative rates (see M. Leclerc and H. H. Britzinger, *J. Am. Chem. Soc.* **117**, 1651–52, 1995).

13. Polymerization of cyclopentene with metallocene catalyst and MAO gives 1,3 rather than 1,2 linked polymers. Why?

Ans. Isomerization precedes insertion (see H. H. Britzinger et al., *Angew. Chem., Int. Ed. Engl.* **34**, 1143–70, 1995; S. W. Collins and W. M. Kelly, *Macromolecules* **25**, 233–37, 1992).

14. "In C_s-symmetric metallocene catalysts syndiospecificity arises due to α-agostic interaction." Discuss the correctness of this statement.

Ans. α-agostic in the transition state and γ-agostic in an intermediate as shown 6.12 and 6.13 are involved. This ensures regular alteration of alkene coordination site and face (see R. H. Grubbs and G. W. Coates *Acc. Chem. Res.* **29**, 85–93, 1996).

15. Attempts to design model complexes of metallocene-based catalytic intermediates have led to the synthesis of $[Cp_2ZrL]^+[MeBPh_3]^-$, where L = 5-methyl, 5-oxo hex-1-ene. Suggest a probable structure.

Ans. Chelation through oxygen and alkene functionality (see Z. Wu, R. F. Jordan, and J. L. Petersen, *J. Am. Chem. Soc.* **117**, 5867, 1995).

16. For metallocene catalysts of general formula $[H_2SiCp_2M\text{-}CH_3]^+$ (M = Ti, Zr, Hf) the relative order of alkene insertion leading to polymerization is Zr > Hf > Ti. What sort of investigation may be required to rationalize this?

Ans. Ab initio theoretical studies that should show activation energies in the reverse order (see T. Yoshida et al., *Organometallics* **14**, 746, 1995).

17. What reaction product is expected when $Cp_2M(CH_3)_2$ (M = Zr, Hf) is treated with BPh_3? Calorimetric measurements show ΔH for these reactions are: −24 (Zr), −21 (Hf). Based on this, what could be said about M–C bond energies?

Ans. Abstraction of CH_3 group, i.e., $[Cp_2M(CH_3)]^+[CH_3BPh_3]^-$. Hf–C stronger than Zr–C (see P. A. Deck and T. J. Marks, *J. Am. Chem. Soc.* **117**, 6128–29, 1995).

18. As mentioned in Section 6.5.3, the use of 6.35/6.36 as a catalyst gives elastomeric polypropylene, i.e., polypropylene with alternate "isotactic" and "atactic" blocks as shown in Fig. 6.10. What effect would bridging the two indenyl ligands with an Me$_2$Si group have on the resultant polymer?

Ans. Equilibrium between racemic 6.35 and meso 6.36 not possible. Consequently with racemic only isotactic and with meso only atactic polymer will be obtained (see J. L. Maciejewshi Petoff et al., *J. Am. Chem. Soc.* **120**, 11316–22, 1998).

19. Alternating ethylene, propylene co-polymers could be obtained by the hydrogenation of polyisoprene. Starting with a mixture of ethylene and propylene, could such a polymer be made by using metallocene catalysts?

Ans. Yes (see R. M. Waymouth and M. K. Leclerc, *Angew. Chem. Int. Ed.* **37**, 922–25, 1998).

20. What could be a research approach for developing catalysts beyond metallocenes?

Ans. See A. M. A. Bennett, *Chemtech*, **29**(7), 24–28, 1999.

BIBLIOGRAPHY

For all the sections

Books

Most of the books listed under Sections 1.3 and 2.1–2.3.4 contain information on polymerization reactions and should be consulted. Specially useful are the book by Parshall and Ittel and Section 2.3.1 of Vol. 1 of *Applied Homogeneous Catalysis with Organometallic Compounds*, ed. by B. Cornils and W. A. Herrmann, VCH, Weinheim, New York, 1996.

For early work on Ziegler–Natta catalysts, see *Ziegler–Natta Catalysts and Polymerizations*, by J. Boor, Jr., Academic Press, New York, 1979.

For background chemistry of sandwich complexes, see *Metallocenes*, by N. J. Long, Oxford: Blackwell Science, 1998.

Also see *Metallocenes: Synthesis, Reactivity, Applications*, ed. by A. Togni and R. L. Halterman, Vols. 1 and 2, Wiley-VCH, New York, 1998.

For properties, processing, and markets of polymers with metallocene catalysts, see *Metallocene-Catalyzed Polymers*, ed. by G. M. Benedikt and B. L. Goodall, Plastic Design Library (William Andrew, Inc.), Norwich, New York, 1998.

For Section 6.7 see the book by Parshall and Ittel.

For Section 6.8 see Kirk and Othmer, *Encyclopedia of Chemical Technology*, 4th edn, John Wiley & Sons, New York, Vol. 6, 1992, pp. 388–458.

Articles

Section 6.1 to 6.3

H. Sinn and W. Kaminsky, in *Advances in Organometallic Chemistry*, ed. by F. G. A. Stone and R. West, Academic Press, Vol. 18, 1980, pp. 99–149.

M. P. McDaniel, in *Advances in Catalysis*, ed. by D. D. Eley, H. Pines, and P. B. Weisz, Academic Press, New York, Vol. 33, 1985, pp. 48–98.

B. L. Goodall, *J. Chem. Edu.* **63**, 191–905, 1986.

W. Keim et al., in *Comprehensive Organometallic Chemistry*, ed. by G. Wilkinson, F. G. A. Stone, E. W. Abel, Pergamon Press, Vol. 8, 1982, pp. 371–462.

G. G. Arzoumanidis and N. M. Karayannis, *Chemtech*, **23**(7), 43–48, 1993.

G. W. Coates and R. M. Waymouth, in *Comprehensive Organometallic Chemistry*, ed. by G. Wilkinson, F. G. A. Stone, and E. W. Abel, Pergamon Press, Vol. 12, 1995, pp. 1193–208.

Section 6.4.1 and 6.4.2

R. H. Grubbs and G. W. Coates, *Acc. Chem. Res.* **29**, 85–93, 1996.

H. H. Britzinger et al., *Angew. Chem. Int. Ed.* **34**, 1143–70, 1995.

Section 6.4.3

P. L. Watson and G. W. Parshall, *Acc. Chem. Res.* **18**, 51–56, 1985.

Section 6.5.1 and 6.5.2

S. S. Reddy and S. Sivaram, *Progr. Polym. Sci.* **20**(2), 309–67, 1995.

K. Richardson, *Chem. Brit.* **30**(2), 87–88, 1994.

C. Janiak, in *Metallocenes: Synthesis, Reactivity, Applications*, ed. by A. Togni and R. L. Halterman, Vols. 1 and 2, Wiley-VCH, New York, 1998, pp. 547–623.

Section 6.5.3

The references listed under Sections 6.4.1 and 6.4.2. Also see G. W. Coates and R. M. Waymouth, *Science*, **67**(5195), 217–19, 1995.

J. A. Ewen et al., *Makromol. Chem. Makromol Symp.* **48/49**, 253–95, 1991.

A. Razavi and J. L. Atwood, *J. Organometchem.* **497**, 105–11, 1995.

For reactions 6.7 and 6.6, see L. Jia et al., *J. Am. Chem. Soc.* **118**, 7900–13, 1996.

G. W. Coates and R. M. Waymouth, *J. Am. Chem. Soc.* **115**, 91–98, 1993.

Also see the references given in answers to Problems 15–19.

Section 6.6

K. H. Theopold, *Chemtech.* **27**(10), 26–32, 1997.

CHAPTER 7

OTHER ALKENE-BASED HOMOGENEOUS CATALYTIC REACTIONS

7.1 INTRODUCTION

In the previous chapters we discussed alkene-based homogeneous catalytic reactions such as hydrocarboxylation, hydroformylation, and polymerization. In this chapter we discuss a number of other homogeneous catalytic reactions where an alkene is one of the basic raw materials. The reactions that fall under this category are many. Some of the industrially important ones are isomerization, hydrogenation, di-, tri-, and oligomerization, metathesis, hydrocyanation, hydrosilylation, C–C coupling, and cyclopropanation. We have encountered most of the basic mechanistic steps involved in these reactions before. Insertions, carbenes, metallocycles, and η^3-allyl complexes are especially important for some of the reactions that we are about to discuss.

7.2 ISOMERIZATION OF ALKENES

Large-scale manufacturing processes involving isomerization reactions by homogeneous catalysts are few. Two important ones are the isomerization step in the SHOP process and the enantioselective isomerization of diethylgeranyl or diethylnerylamine as practiced by the Takasago Perfumery. The isomerization step in the SHOP process is discussed later on in this chapter. The Takasago process is discussed in Chapter 9. Isomerization of alkenes with nitrile functionalities is very important in DuPont's hydrocyanation process. The mechanism for this reaction is discussed in Section 7.7.

7.2.1 Catalytic Cycle

In any homogeneous catalytic reaction involving an alkene, isomerization of the alkene is always a possibility. The insertion of alkene into the M–H bond can occur in a Markovnikov or anti-Markovnikov manner (see Section 5.2.2). Alkene isomerization involves Markovnikov addition, which is followed by a β-hydride elimination. A simplified catalytic cycle is shown in Fig. 7.1.

In Fig. 7.1, the metal hydride 7.1 is the important catalytic intermediate. Conversion of 7.2 to 7.3 involves insertion of 1-butene in the M–H bond in a Markovnikov manner. Complex 7.4 is the anti-Markovnikov product. But-2-ene is formed from the Markovnikov product by β-hydride elimination. A slightly different mechanism involves 1,3-hydrogen shift and the involvement of a η^3-allyl rather than a metal–alkyl intermediate. This is shown schematically in Fig. 7.2.

The basic difference between the two mechanisms is that the oxidation state of the metal ion changes by 2+ in going from 7.6 to 7.7, and by 2− in going from 7.7 to 7.8. In contrast, for the mechanism outlined in Fig. 7.1 all the catalytic intermediates have the metal atom in the same oxidation state. The switch over from 7.6 to 7.7 can be formally worked out by the familiar "electron-pair pushing" principle of organic chemistry. This is shown by 7.1. Note that formally an electron pair is transferred from the metal to the ligand, causing its oxidation state to increase by two.

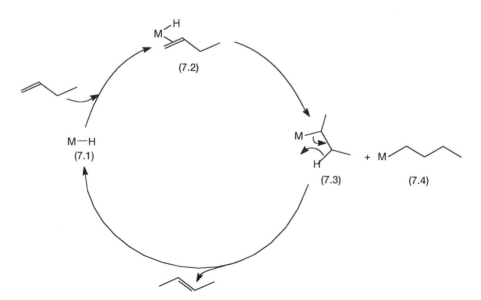

Figure 7.1 Isomerization by the "hydride" mechanism. Note that β-hydride abstraction from the other β-carbon takes 7.3 back to 7.2.

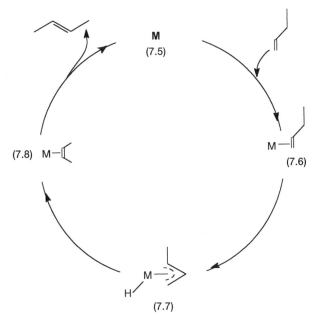

Figure 7.2 Isomerization by the "allyl" mechanism. Note that 7.6 to 7.7 is an oxidative addition reaction, as shown in 7.1.

$$(7.1)$$

7.3 HYDROGENATION OF ALKENES

The industrial importance of homogeneous hydrogenation lies mainly in the fact that by using a chiral catalyst hydrogenated chiral products could be obtained from prochiral alkenes (see Table 9.1). With achiral catalysts also, a very high degree of chemo- and regioselectivity could be obtained. An example of this is shown by reaction (7.2). As could be seen, the disubstituted rather than the trisubstituted alkene functionality is preferentially hydrogenated. Hydrogenation of alkenes is one of the most well-studied homogeneous catalytic reactions.

$$(7.2)$$

In the following section, we discuss the basic mechanism of homogeneous hydrogenation by Wilkinson's catalyst, $RhCl(PPh_3)_3$. Many other complexes of rhodium as well as complexes of other metals such as ruthenium, platinum, lutetium, etc. have also been used as homogeneous, laboratory-scale, hydrogenation catalysts. The mechanisms in all these cases may differ substantially.

7.3.1   Catalytic Cycle

The basic mechanism of hydrogenation is shown by the catalytic cycle in Fig. 7.3. This cycle is simplified, and some reactions are not shown. Intermediate 7.9 is a 14-electron complex (see Section 2.1). Phosphine dissociation of Wilkinson's complex leads to its formation. Conversion of 7.9 to 7.10 is a simple oxidative addition of H_2 to the former. Coordination by the alkene, for example, 1-butene, generates 7.11. Subsequent insertion of the alkene into the metal–hydrogen bond gives the metal alkyl species 7.12. The latter undergoes reductive elimination of butane and regenerates 7.9.

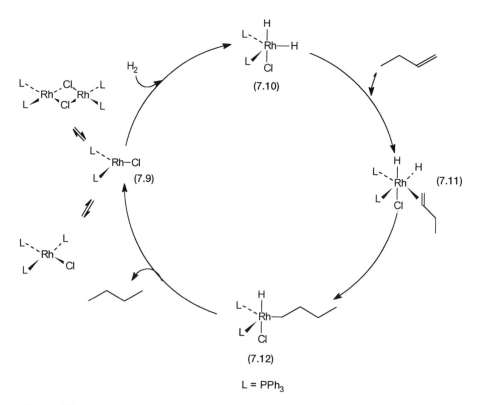

Figure 7.3   Hydrogenation by Wilkinson's catalyst. For butene, oxidative addition of H_2 precedes alkene coordination but not necessarily for all other alkenes.

Note that although the conversion of 7.11 to 7.12 assumes anti-Markovnikov addition, the Markovnikov product also gives butane. Conversion of 7.9 to 7.11 could also take place by prior coordination of alkene followed by the oxidative addition of dihydrogen. Indeed this parallel pathway for the formation of 7.11 does operate. Like the equilibrium shown between $RhClL_3$, 7.9, and the dimer $[RhClL_2]_2$, there is an equilibrium between 7.9 and the alkene coordinated complex $RhCl(alkene)L_2$.

7.3.2 Mechanistic Evidence

Homogeneous hydrogenation by $RhClL_3$ has many instructive points about the successful use of kinetic methods for the elucidation of mechanism. Under a wide range of conditions, the rate law is found to obey the empirical relationship as shown by 7.3. A little algebraic manipulation and rearrangement would show that if the other variables are kept constant, then rate is inversely proportional to the concentration of L. This, of course, is consistent with the generation of 7.9 from Wilkinson's complex by phosphine dissociation. The rate law also shows saturation kinetics with respect to both dihydrogen and alkene concentrations (Section 2.5.1).

$$r = [Rh]\left(a + \frac{b[L]}{[H_2]} + \frac{c[L]}{[S]}\right)^{-1}$$

$$S = alkene \quad r = rate = -d[S]/dt$$

$$(7.3)$$

The physical significance of the rate law is that consumption of alkene occurs by two pathways. One involves a rapid pre-equilibrium between 7.9 and dihydrogen to give 7.10. This is followed by a reaction with the alkene that eventually leads to the formation of the alkane. The other involves a similar equilibrium between 7.9 and alkene to give an alkene complex. This alkene complex then reacts with dihydrogen. In Fig. 7.3 for simplicity only the forward reaction of the former equilibrium has been shown.

Although the mechanism shown in Fig. 7.3 broadly applies to a variety of alkenes, the details and the sequence of reactions could vary significantly. We will see that for the hydrogenation of α-acetamidocinnamic acid, the catalyst–alkene complex is formed first, which then oxidatively adds to dihydrogen. More precisely, out of the two equilibria, one with the alkene and the other with dihydrogen, which one would dominate is determined by the nature of the alkene. Even within the class of unfunctionalized alkenes there may be significant differences between the types of catalytic intermediates that are involved.

Reaction of 7.9 with dihydrogen has been studied by NMR spectroscopy. Two isomers of 7.10, one as shown in Fig. 7.3 and the other where the two Ls are *trans* and occupy axial positions (Complex 7.13 in Fig. 7.4), have been

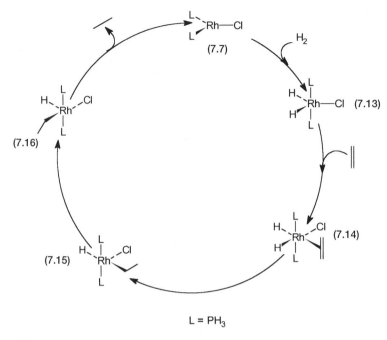

L = PH₃

Figure 7.4 Catalytic cycle assumed for the ab initio theoretical calculations of alkene hydrogenation. Note that structure 7.10 is different from 7.13, and in solution these are in rapid equilibrium.

observed. The energy required for the interconversion between these two isomers is of course small. As we have seen earlier (Sections 2.3.3 and 2.5.3) both insertion and reductive elimination have been well studied in model systems.

Ab initio quantum-mechanical calculations have also been carried out to estimate theoretically the relative energies of different transition states and intermediates. To keep the calculations at a manageable level the model catalytic cycle shown in Fig. 7.4 has been considered. According to this calculation, conversion of 7.14 to 7.15 is found to be the rate-determining step. This is consistent with the empirical kinetic results. Isomerization of 7.15 to 7.16 is found to be thermodynamically highly favorable. The reductive elimination step, that is, 7.16 to 7.7, is found to have a very low free energy of activation and be thermoneutral, that is, thermodynamically neither strongly favorable nor unfavorable.

7.4 OLIGOMERIZATION OF ETHYLENE

Oligomerization of ethylene to give linear terminal alkenes or the so-called α-olefins is a very important industrial reaction. The linear α-olefins with 10 to

18 carbon atoms are important feed stock for a variety of detergents. Oligomers with 4–10 carbon atoms find uses as co-monomers in the manufacture of polyethylene and also as starting materials for the manufacture of plasticizer alcohols by the hydroformylation reaction.

Industrial processes utilize aluminum or nickel complexes as catalysts for oligomerization. Both Gulf Oil and Ethyl Corporation use aluminum alkyls, while the Shell higher olefin process (SHOP) involves the use of a nickel catalyst. Both these basic reactions (i.e., oligomerization of ethylene by aluminum alkyl, "Aufbaurektion," and the fact that in the presence of nickel such an oligomerization reaction yields mainly butene (the "nickel effect"), were discovered by Ziegler in the early 1950s. The mechanism of oligomerization is basically the same as that of ethylene polymerization (see Section 6.4.1). However, in this case chain termination by β-elimination has to occur far more frequently to keep the chain length down.

Stoichiometric reaction of the type shown by 7.4 also leads to the formation of ethylene oligomers. In the Ethyl Corporation process one step involves stoichiometric reaction of this type. Another variant of this is the Conco process, where such stoichiometric reactions are followed by oxidation and hydrolysis of the aluminum alkyls. This gives linear α-alcohols that are used in biodegradable detergents. The co-product is highly pure alumina, which has a variety of uses, including that of an acidic heterogeneous catalyst.

$$Al(C_2H_5)_3 + 3n\ C_2H_4 \longrightarrow Al[(CH_2)_{2n+1}\ CH_3]_3$$

$$(7.4)$$

7.4.1 Shell Higher Olefin Process

SHOP involves essentially four sequential operations. First ethylene is oligomerized with a soluble nickel catalyst to give linear α-alkenes. In the second step, over a heterogeneous catalyst these are isomerized to internal alkenes. In the third step internal alkenes containing four to eight carbon atoms are mixed with internal alkenes of 20 and more carbon atoms and the mixture is subjected to the metathesis reaction (see Section 7.6). In the fourth step three reactions —isomerization, hydroformylation, and hydrogenation—are simultaneously carried out with a homogeneous cobalt catalyst to give long-chain α-alcohols. The reactions of the fourth step were discussed in Chapter 5 (see Table 5.1 and Section 5.3). The sequence of reactions is schematically shown in Fig. 7.5. In this section, we discuss only the first step, the oligomerization of ethylene to give linear α-olefins by soluble nickel catalysts.

As already mentioned, the mechanism of oligomerization is the same as discussed for polymerization, and a catalytic cycle similar to the one shown in Fig. 6.3 operates. Many nickel–phosphine complexes have been successfully used as the precatalysts; 7.17 is one such example. As shown by 7.5, reaction of a phosphorous ylide with a suitable nickel-containing precursor makes this

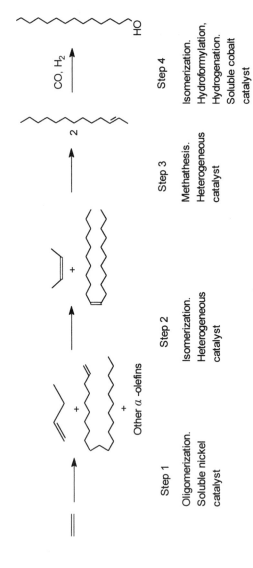

Figure 7.5 The four steps of Shell's higher olefin process (SHOP). Four- and 22-carbon-atom alkenes are taken as representative examples.

complex. The selectivity of the oligomerization reaction is largely controlled by the chelating ligand.

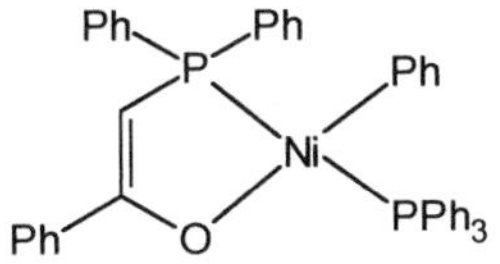

$$(7.17)$$

$$(7.5)$$

Under the reaction conditions the precursor complex probably generates a nickel–hydride species, which then initiates the oligomerization reaction. Evidence for this comes from the studies on the reactions of 7.17. As shown by 7.6, on reaction with ethylene 7.17 eliminates styrene and produces a nickel–hydride complex. A model catalytic intermediate 7.18 has been characterized by single-crystal X-ray studies. Complex 7.18 reacts with ethylene to give a nickel–ethyl species in a reversible manner. This is shown by reaction 7.7. Reactions 7.6 and 7.7 are strong evidence for the involvement of a nickel–hydride catalytic intermediate.

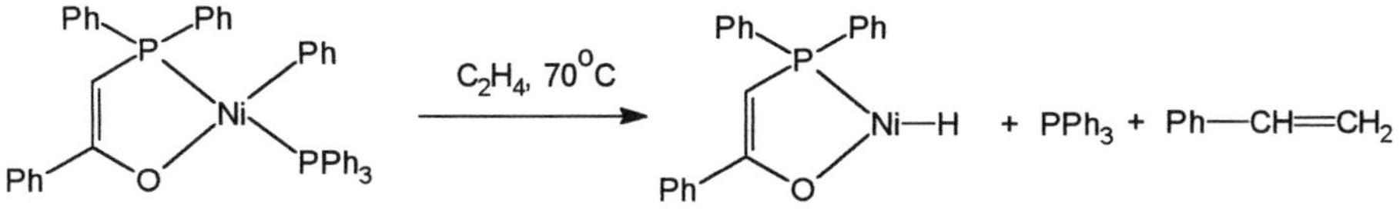

$$(7.6)$$

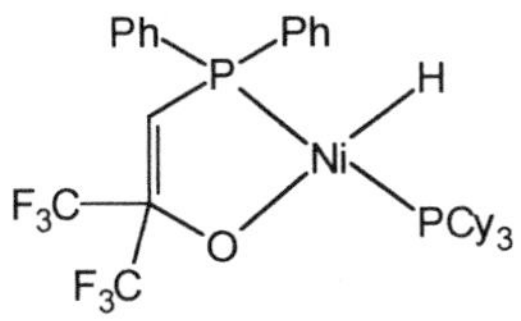

$$(7.18)$$

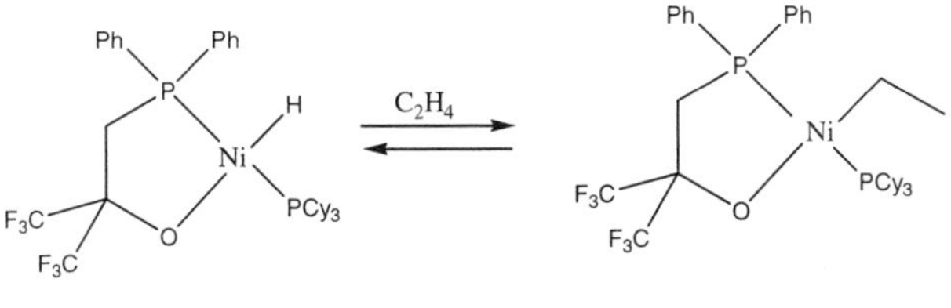

$$(7.7)$$

7.5 DI-, TRI-, AND CODIMERIZATION REACTIONS

One of the industrially important dimerization reactions that involves the use of homogeneous catalysts is the dimerization of propylene. Dimerization of propylene produces mixtures of the isomers of methyl pentenes, hexenes, and 2,3-dimethyl butene and is practiced by the Institut Francis du Petrole (IFP), Sumitomo, and British Petroleum (BP). The methyl pentenes and hexenes are used as gasoline additives. Dimethylbutene is used in the fragrance and the agrochemical industries.

Dimerization of butadiene is used for the selective formation of 1,5-cyclo-octadiene (1,5-COD), which on selective hydrogenation gives cyclooctene. By ring-opened metathesis polymerization of cyclooctene a specialty polymer is obtained (see Section 7.6.1). Hulls sells this polymer as Vestenamer®.

Butadiene could also be trimerized to give cyclododecatriene. The trimer is again used by Hulls to manufacture nylon 12 and Vestamid®. The codimerization of butadiene and ethylene is used by DuPont to manufacture 1,4-hexadiene, one of the monomers of EPDM (ethylene, propylene, diene, monomers) rubber. The role of the diene monomer in EPDM rubber is to provide with two double bonds of different reactivities. The more reactive, terminal double bond takes part in the polymerization with ethylene and propylene. The less reactive internal one is used later on for cross-linking. These important catalytic reactions are shown in Fig. 7.6.

7.5.1 Dimerization of Propylene

The dimerization of propylene carried out by IFP is called the DIMEROSOL process and involves the use of nickel catalysts. This is shown in Fig. 7.7. Complexes 7.20 and 7.21 are the anti-Markovnikov and Markovnikov insertion products into the Ni–H bond. Structures 7.23(A) and (B) are intermediates derived from 7.21 by inserting the *second* propylene molecule in a Markovnikov and anti-Markovnikov manner, respectively. Similarly 7.22(A) and (B) are intermediates from 7.20 by the insertion of the second propylene molecule. These four nickel–alkyl intermediates by β-elimination give six alkenes. Under the process conditions these alkenes may undergo further isomerization.

7.5.2 Di- and Trimerization of Butadiene

Both di- and trimerization of butadiene with soluble nickel catalysts are well-established homogeneous catalytic reactions. The precatalyst having nickel in the zero oxidation state may be generated in many ways. Reduction of a Ni^{2+} salt or a coordination complex such as $Ni(acac)_2$ (acac = acetylacetonate) with alkyl aluminum reagent in the presence of butadiene and a suitable tertiary phosphine is the preferred method. The nature of the phosphine ligand plays an important role in determining both the activity and selectivity of the catalytic

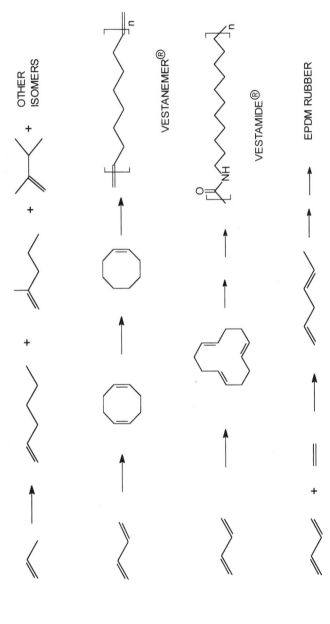

Figure 7.6 Industrial use of (from the top): propylene dimerization, butadiene dimerization, butadiene trimerization, and butadiene plus ethylene codimerization. In EPDM rubber, the terminal double bond of 1,4-hexadiene takes part in polymer formation. The internal double bond is used during curing.

143

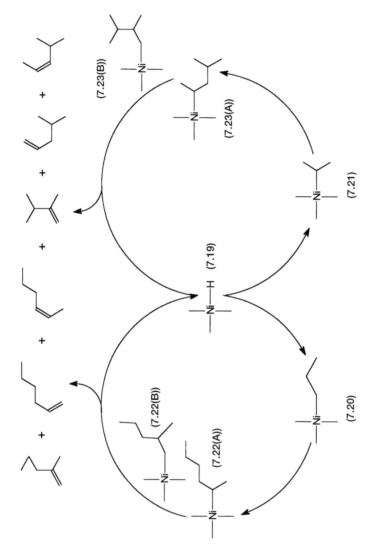

Figure 7.7 Catalytic cycles for propylene dimerization. For the first propylene molecule the left and right cycles represent the anti-Markovnikov and Markovnikov pathways, respectively. Note that for 7.22(A) and 7.23(A) there are two β-hydride positions.

system. The other products that may be formed in this reaction are vinylcy-clohexene and divinylcyclobutane. By an optimal choice of ligand and reaction conditions, formation of these by-products may be avoided to a large extent, and 1,5-cyclooctadiene may be obtained with >96% selectivity.

The catalytic cycle proposed for the dimerization of butadiene is shown in Fig. 7.8. As shown by 7.24, two molecules of butadiene coordinate to NiL. A formal oxidative addition, as shown by Eq. 7.8, produces two nickel–carbon bonds and the carbon–carbon bond required for ring formation. The structure of 7.25 with two nickel–carbon bonds (see Fig. 7.8), is a hypothetical one that helps us to understand the carbon–carbon bond formation process. The actual catalytic intermediates that have been observed by spectroscopy have an η^3-allyl type of bonding. As shown by reaction 7.9, species 7.25 can reductively eliminate 1,5-cyclooctadiene and the zerovalent nickel complex Ni–L.

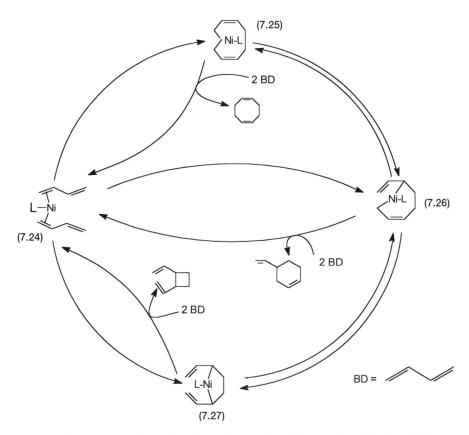

Figure 7.8 Dimerization of butadiene. L is a phosphine or phosphite (see Problem 11). 7.26 and 7.27 can be formed from butadiene (BD) and L–Ni by 7.8-type formalism.

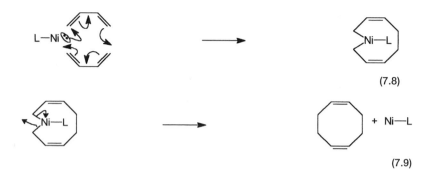

(7.8)

(7.9)

Similar oxidative additions involving the inner carbon atoms of the butadiene molecules can generate complexes having the formal structures 7.26 and 7.27. These may also be formed from 7.25 through tautomerization. Regeneration of 7.24 from these species involves elimination of vinyl cyclohexene and divinyl cyclobutane, respectively.

The evidence for the proposed mechanism as shown in Fig. 7.8 comes mainly from in situ NMR studies and X-ray structures of isolated model complexes. Rapid equilibrium between species 7.25 to 7.27 involving an η^3-allyl type of interaction results in a species of the type 7.28. This species has been observed by NMR. Similar model complexes such as 7.29 and 7.30 have been characterized by single crystal X-ray studies.

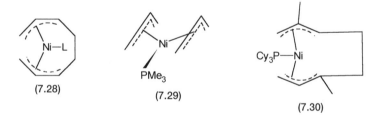

(7.28)

(7.29)

(7.30)

In the absence of added phosphine the main product is the cyclic *trimer* of butadiene—cyclododecatriene. The presence of three double bonds in this molecule means other geometric isomers apart from the one shown in Fig. 7.6 exist. Identification of the species 7.31 by NMR is evidence for the involvement of η^3-allyl intermediates. The complex 7.31 reductively eliminates cyclododecatriene.

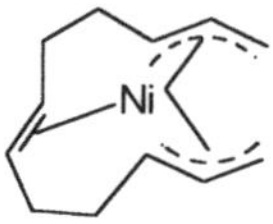

(7.31)

7.5.3 Dimerization of Butadiene with Ethylene

Butadiene and ethylene are codimerized with a soluble rhodium–phosphine complex as the catalyst. Very little has been reported on the mechanistic evidence for this reaction. However, a catalytic cycle as shown in Fig. 7.9 involving a rhodium hydride seems likely. Reducing rhodium trichloride with ethanol in the presence of a tertiary phosphine generates the hydride complex 7.32. The 1,4-hydride attack on the coordinated butadiene gives an η^3-allyl complex. This is shown by the conversion of 7.33 to 7.34. Ethylene coordination to 7.34 produces 7.35.

The latter undergoes insertion of ethylene into the rhodium–carbon σ bond to give 7.36. The formal mechanism for the formation of 7.36 is shown by reaction 7.10. Complex 7.36 undergoes β-elimination to generate 7.37, which liberates 1,4-hexadiene and completes the catalytic cycle.

$$(7.10)$$

7.6 METATHESIS REACTIONS

Metathesis of alkenes is essentially a class of reactions where an interchange of C atoms between pairs of double bonds takes place. A few representative examples are shown by the reactions listed in Fig. 7.10. The industrial use of metathesis reactions so far has been limited mainly to exchange metathesis (Fig. 7.10, top, backward reaction) as in the SHOP process, and ring-open metathesis polymerization (ROMP). As already mentioned (Section 7.5), Vastenamer® is a polymer made by Hulls by ROMP from cyclooctene. Similarly, the polymer from norborene by ROMP is manufactured by CdF Chemie and is sold by the trade name of Norsorex®.

Almost all industrial processes based on metathesis reactions involve heterogeneous catalysts. These are made by supporting molybdenum, tungsten, or rhenium in a high oxidation state on an inorganic support like alumina. A liquid-phase process involving a Ziegler type of catalytic system based on ruthenium has recently been reported by Hercules, for the manufacture of a specialty polymer from dicyclopentadiene. The importance of the homogeneous catalytic systems as practiced at the laboratory scale is that they provide a comprehensive understanding of the metathesis process at a molecular level.

7.6.1 Mechanistic Studies

The mechanism of all metathesis reactions consists of three basic steps. The first step is the formation of metal–alkylidene complexes. The second step is the formation of metallocyclobutanes. The third step is the opening of the metallocyclobutane rings, which leads to product formation. The catalytic cycle

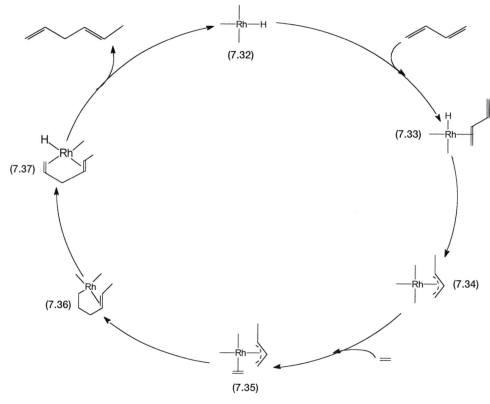

Figure 7.9   Dimerization of butadiene and ethylene by a rhodium–hydride catalyst. Note the allyl bonding proposed in 7.34 and 7.35.

shown in Fig. 7.11 consists of these three basic steps. The alkylidene species 7.38 or 7.40 is formed by the reactions of the precatalyst M, with the two hypothetical alkenes.

The most likely way that this can happen is through the formation of metal– alkyls followed by α-elimination, as shown by reaction 7.11. The primary al- kylidiene species thus generated can undergo reactions 7.12 and 7.13. The alkenes that are generated in reaction 7.12 and 7.13 are obviously not the products of the catalytic cycle; they are generated only initially when the pre- catalyst is converted into the active catalytic intermediates. Note that similar pathways involving the other alkene also exist. These are not shown.

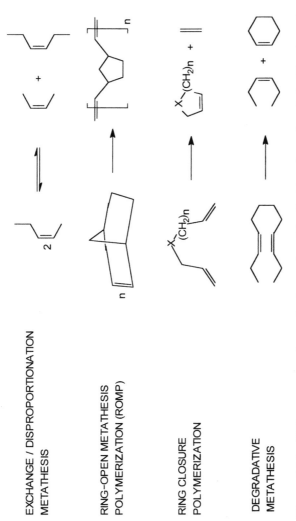

EXCHANGE / DISPROPORTIONATION
METATHESIS

RING-OPEN METATHESIS
POLYMERIZATION (ROMP)

RING CLOSURE
POLYMERIZATION

DEGRADATIVE
METATHESIS

Figure 7.10 Different types of metathesis reaction. The first two have found significant use in industry.

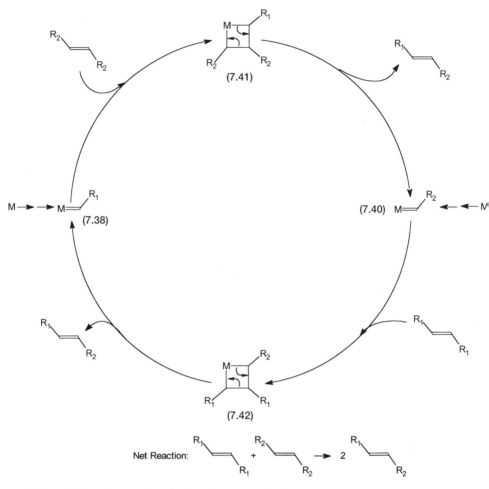

Figure 7.11   A general cycle for alkene metathesis. Note that both the carbenes 7.38 and 7.40 can catalyze the metathesis reaction.

The evidence for the proposed mechanism and reactions 7.11 to 7.13 come from a variety of observations. First of all cleavage of the alkenes only at the double bonds, that is, generation of species such as 7.38 and 7.40, is indicated by isotope-labeling studies. A mixture of but-2-ene and perdeuterated but-2-ene on exposure to metathesis catalysts shows that the product but-2-ene is duterated only at the 1,2 positions. Second, fully characterized metal–alkylidene complexes such as 7.43 and 7.44 have been shown to be active metathesis catalysts.

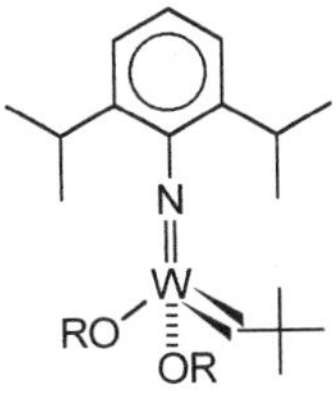

(7.43)

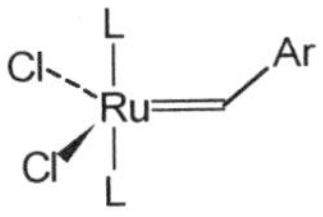

(7.44)

The product from reaction 7.14 has been isolated and fully characterized. This lends credence to the mechanistic postulate that metallocyclobutane rings play a crucial role in metathesis reactions.

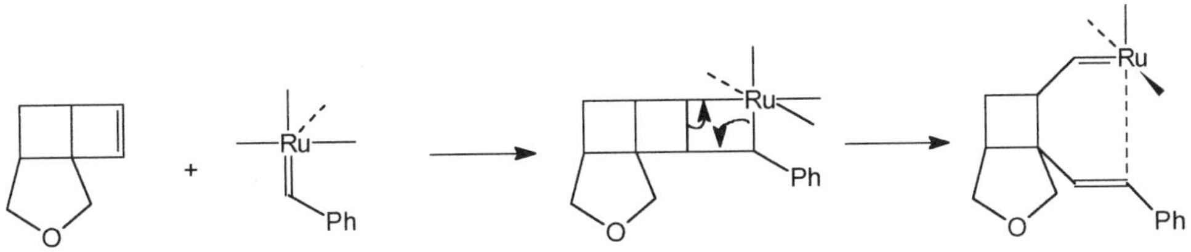

(7.14)

In ROMP reactions when the metal–alkylidene species remain attached to the polymer chain, the polymerization process is called *living polymerization*. Such a situation is possible if the extent of side reactions (e.g., exchange metathesis, etc.) is minimal. The mechanism by which chain propagation takes place in the ROMP reaction of an alkene such as cyclooctene is shown by Fig. 7.12. It should be noted that any metal alkylidene, not necessarily the one derived from cyclooctene, could initiate and propagate the ROMP reaction. In the case of living polymerization the end group of the polymer chain would, however, be identical to the organic fragment of the alkylidene species that initiates the polymerization.

7.7 HYDROCYANATION

DuPont manufactures adiponitrile (ADN), a raw material for nylon 6,6, by the hydrocyanation of butadiene using homogeneous nickel catalysts. As shown

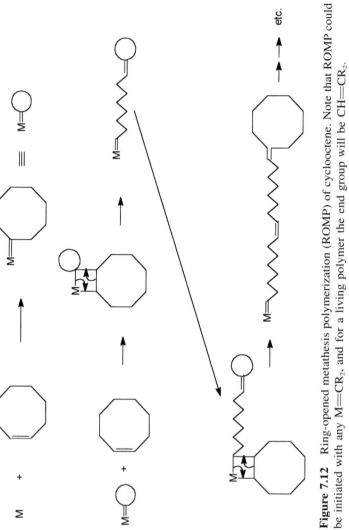

Figure 7.12 Ring-opened metathesis polymerization (ROMP) of cyclooctene. Note that ROMP could be initiated with any M=CR$_2$, and for a living polymer the end group will be CH=CR$_2$.

by reaction 7.15, this involves the addition of two molecules of HCN to butadiene.

(7.15)

The reaction is carried out in two stages. In the first stage one molecule of HCN is added to butadiene to give 3-pentenenitrile (3PN) and 2-methyl 3-butenenitrile (2M3BN) by anti-Markovnikov and Markovnikov addition of the CN^- group, respectively. Under the reaction conditions, as shown by reaction 7.17, 2M3BN is isomerized to 3PN. So the first stage involves reactions 7.16 and 7.17. In the second stage 3PN is isomerized to 4-pentenenitrile and the second molecule of HCN is added to 4-pentenenitrile to give ADN, the desired product. Reactions 7.18 and 7.19 show these steps.

First Stage:

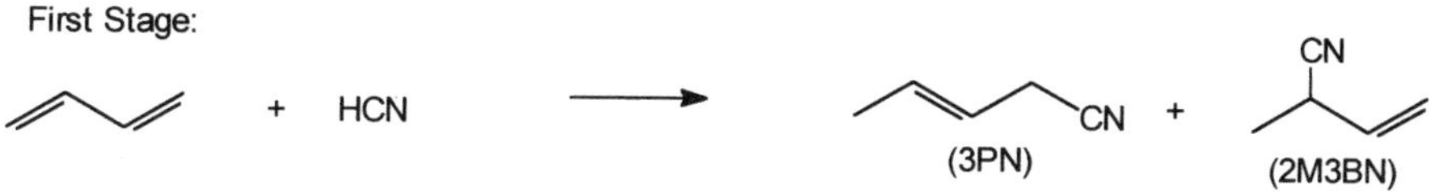

(7.16)

(7.17)

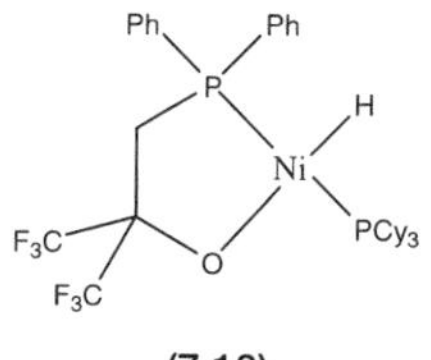

(7.18)

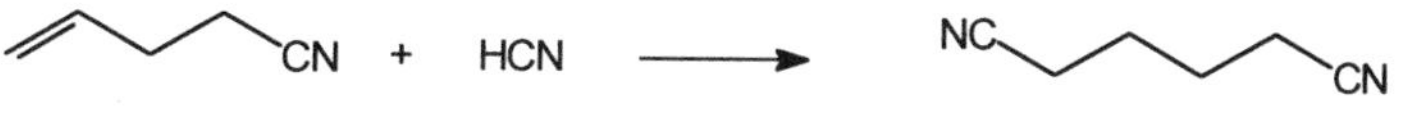

(7.19)

The hydrocyanation reaction is important not only because it is practiced industrially on a large scale, but also because it clearly illustrates some of the fundamental postulates of homogeneous catalysis. The potential of the hydrocyanation reaction in asymmetric catalysis has also been explored and appears

7.7.1 Catalysts for Hydrocyanation

All the reactions of the hydrocyanation process are catalyzed by zero-valent nickel phosphine or phosphite complexes. These are used in combination with Lewis acid promoters such as zinc chloride, trialkyl boron compounds, or trialkyl borate ester. The ability of the precatalyst to undergo ligand dissociation followed by oxidative addition of HCN plays a crucial role in the hydrocyanation reaction. The Lewis acid is very important for the facile and selective formation of 3PN and ADN in the first and the second stage, respectively.

The energetics of the ligand dissociation process, reaction 7.20, depends on the electronic and steric characteristics of the ligand. The equilibrium constants for different Ls are found to be correlated to the cone angles of the phosphorus ligand: The larger the cone angle, the bigger the equilibrium constant. However, the rate at which this equilibrium is established (i.e., the kinetics of the process) does not appear to depend on the steric bulk.

$$NiL_4 \rightleftharpoons NiL_3 + L \qquad\qquad (7.20)$$

The essential role of the Lewis acid is to act as an electron acceptor of the nitrogen lone pair of the coordinated cyano group. Such an interaction increases the steric crowding in catalytic intermediates, weakens the Ni–CN bond, and stabilizes the catalytic intermediates with respect to degradation. The net effect is an enhancement in the rates of formation of the desired products in both stages of the hydrocyanation process. The Lewis acid interacts with the coordinated cyano group rather than the nitrile functionality of the free nitrile. This is because of the higher Lewis basicity of the coordinated cyano group.

7.7.2   Catalytic Cycle for the First Stage

The catalytic cycle for the formation of 3PN and 2M3BN is shown in Fig. 7.13. The following points deserve attention. First, interactions of Lewis acid with the coordinated cyano groups are not shown. The evidence for such interactions comes from full characterization of $HNiL_3CN–BPh_3$ (L = o-tolylphosphite) as well as detailed IR and multinuclear NMR studies of a number of analogous nickel complexes. Second, the intermediates NiL_3 and 7.45–7.47 are well characterized by IR and multinuclear NMR techniques. Single crystal X-ray structures of NiL_2 (alkene), where the alkene is ethylene or acrylonitrile and L is o-tolyl phosphite, have been determined. These are obvious models for the proposed intermediates 7.48 and 7.50. As both these intermediates are 16-electron species, addition of L to give 7.49 or 7.51 prior to the elimination of 3PN or 2M3BN are reasonable mechanistic steps.

Although for clarity the reactions of Fig. 7.13 are shown to be unidirectional, all the reactions of the catalytic cycle are in fact reversible. This is an important aspect of the first stage of the hydrocyanation process. It provides for a mechanism for the isomerization of the unwanted 2M3BN to the desired 3PN. The isomerization reaction of 2M3BN to 3PN has been studied by deuterium-labeling experiments. The results are consistent with a mechanism where butadiene is formed in one of the intermediate steps. This means that the reversibility of all the steps allows isomerization to follow the path: 7.51 → 7.50 → 7.47 → 7.46 → 7.47 → 7.48 → 7.49.

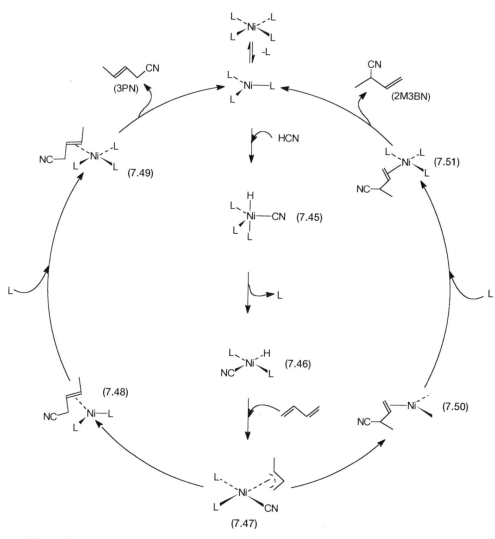

Figure 7.13 First stage of hydrocyanation. Conversion of butadiene to 3PN. Under the reaction conditions 2M3BN is isomerized to 3PN. Interaction of Lewis acid with coordinated nitrile is not shown for clarity. The left and right side involve CN⁻ addition in an anti-Markovnikov and Markovnikov manner. L = P(OEt)₃ or P(O-*o*-tolyl)₃.

Note that dehydrocyanation of 2M3BN (i.e., 7.50 to 7.47) is nothing but a simple oxidative addition reaction. This is shown formally by reaction 7.21. Reaction 7.22 shows the formal mechanism of butadiene formation and conversion of 7.47 to 7.46. Between the two isomers, 3PN is thermodynamically more stable than 2M3BN. A mixture of these two nitriles, if allowed to reach a thermodynamic equilibrium over a catalyst, would have the concentration ratio of approximately 9:1.

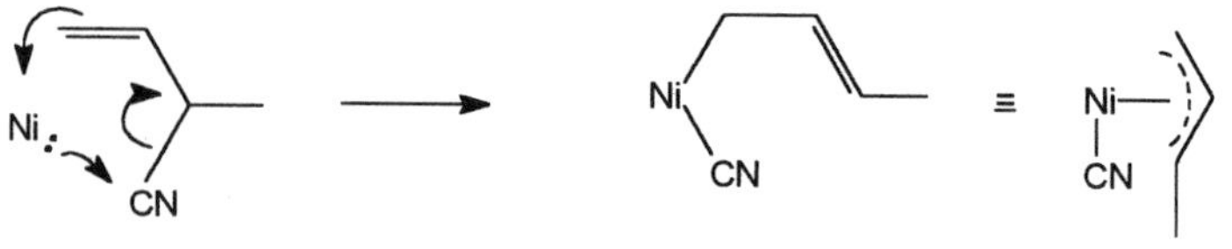

(7.21)

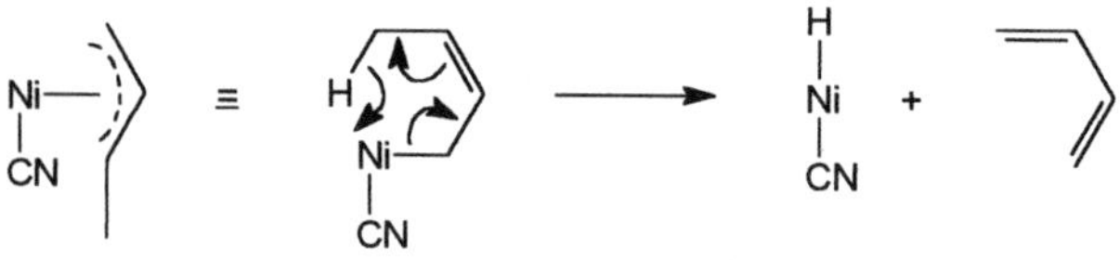

(7.22)

7.7.3   Catalytic Cycle for the Second Stage

As mentioned earlier, the first reaction in the second stage is the isomerization of 3PN to 4PN, reaction 7.18. The mechanism of this reaction is very similar to the mechanism of alkene isomerization discussed in Section 7.2.1, and shown by Fig. 7.14. The following points need attention.

The nickel–hydride complex that acts as a precatalyst for this isomerization reaction is thought to be the cationic part of 7.52. The evidence for the existence and participation of a cationic species such as 7.52 comes from multinuclear NMR and IR data. An equilibrium as shown by 7.23 exists, and the cation $[HNiL_4]^+$ is the dominant precatalyst for the isomerization reaction. The cation is an 18-electron complex. It undergoes ligand dissociation to give 7.53 before alkene coordination takes place. A ligand dissociated species such as 7.53 with L = p-tolylphosphite has been observed spectroscopically at low temperatures.

$$HNiL_3(CN\text{-}A) + L \rightleftharpoons [HNiL_4]^+ [CN\text{-}A]^- \qquad (7.23)$$

The isomerization of 3PN can lead to two possible products, 2-pentenenitrile (2PN), the unwanted isomer, and 4PN, the desired isomer. The former does *not* undergo hydrocyanation and thermodynamically is the most stable isomer. If the isomerization of 3PN were allowed to reach thermodynamic equilibrium, the concentrations of the three isomers 2PN, 3PN, and 4PN would be approximately 78:20:2. Fortunately the isomerization of 3PN to 4PN is about 70 times as fast as that of 3PN to 2PN. In other words, although 4PN is thermodynamically the less stable isomer, the favorable kinetics allows its preferential formation.

The catalytic cycle for the hydrocyanation of 4PN to desired adiponitrile and undesired 2-methyl glutaronitrile (MGN) is shown by Fig. 7.15. The intermediates that lead to the formation of 7.56 or 7.57 from NiL$_3$ are not shown

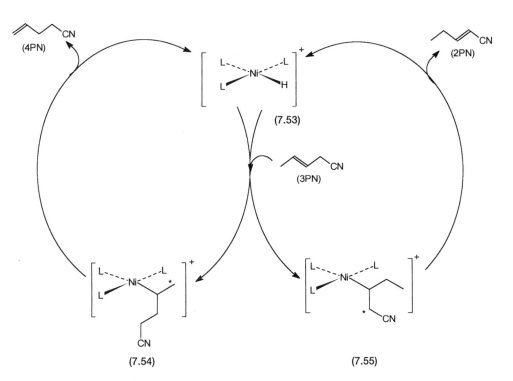

Figure 7.14   Second stage of hydrocyanation. Isomerization of 3PN to 4PN, the desired isomer, and 2PN, the unwanted isomer. The C atoms marked by asterisks are the ones from which the β-hydride abstraction takes place.

but are considered to be 7.45 and 7.46, with the Lewis acid A attached to the coordinated cyano group.

The left- and right-hand loops of the catalytic cycle involves anti-Markovnikov and Markovnikov additions, giving 7.56 and 7.57, respectively. The presence of the Lewis acid ensures that the former pathway is favored. Spectroscopic evidence for species 7.56 or 7.57 have so far not been reported. However, dissociation of two moles of ligand and involvement of the Lewis

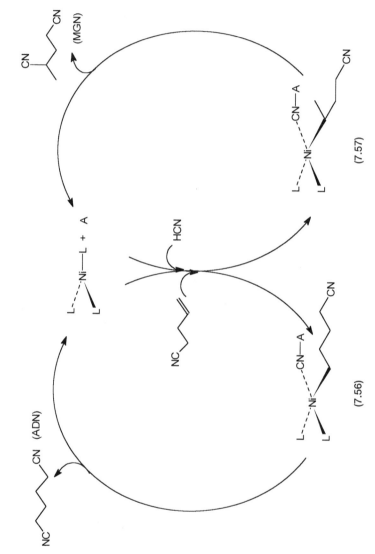

Figure 7.15 Second stage of hydrocyanation. Conversion of 4PN to adiponitrile (ADN) and MGN. A is the Lewis acid. The left loop dominates, giving ADN as the main product. Analogues of 7.45 and 7.46 with Lewis acid are not shown for clarity but are definitely involved.

acid in the transition state are indicated by the empirically determined rate law. This is shown by the following rate expression.

$$\text{Rate} = k \; \frac{[\text{NiL}_4]\,[\text{4PN}]\,[\text{A}]\,[\text{HCN}]}{[\text{L}]^2} \qquad (7.24)$$

$$(\text{A} = \text{LEWIS ACID})$$

7.8 HYDROSILYLATION

The addition of a silicon compound such as R_3SiH to alkene functionality, as shown by reaction 7.25, is used widely in silicone polymer manufacture and is called the hydrosilylation reaction. Although hydrosilylation was discovered in 1947, the first homogeneous catalyst, H_2PtCl_6 (1–10%), in 2-propanol was reported in 1957 from the laboratories of Dow Corning.

$$(7.25)$$

The importance of hydrosilylation reaction in the context of homogeneous catalysis is twofold. First, the hydrosilylation reaction is very important in the "curing" of silicone rubber. Such curing leads to cross-linking of the polymer chains and turns a "syrupy" polymer to a "gum" rubber or a "soft" polymer to a "hard" one. Reactions 7.26 and 7.27 show the formations of new bonds that can bring about these changes in the polymer properties. Hydrosilylation reactions have also been used to give nitrile or a CF_3 group containing monomers. These monomers in turn can be used to give silicone polymers of special properties.

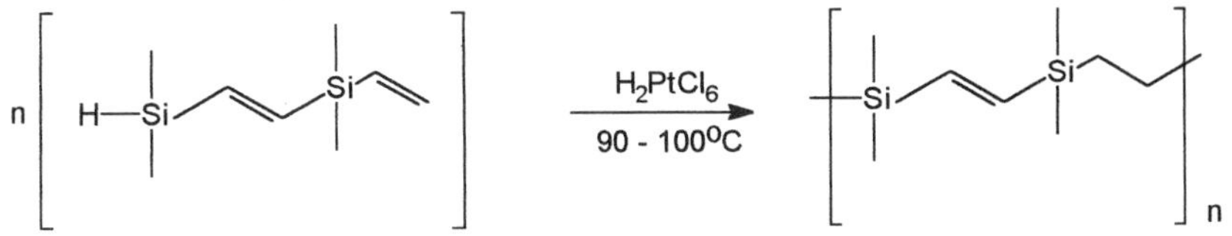

$$(7.26)$$

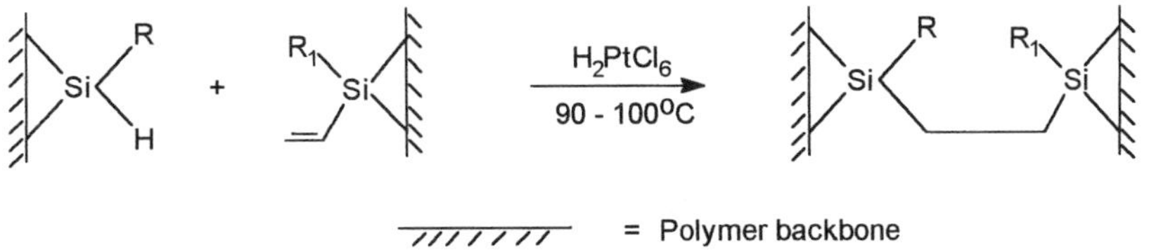

$$\overline{\text{////////}} = \text{Polymer backbone}$$

$$(7.27)$$

Second, the formal dividing line between homogeneous and heterogeneous catalysis becomes blurred in the case of hydrosilylation reactions. In a number of such reactions, recent evidence shows that metal colloids as well as homogeneous catalytic intermediates may be involved. The involvement of metal colloids is well established for all H_2PtCl_6 plus isopropanol-based reactions. Like the Zeigler catalysts, most of these catalytic systems are therefore closer to heterogeneous ones. In fact, even with fully characterized soluble platinum complexes of the type 7.58, under the conditions of the hydrosilylation reaction colloidal platinum is eventually generated. However, there is evidence to show that with these precatalysts mononuclear catalytic intermediates are also involved.

(7.58)

In reaction systems where the ligands prevent colloid formation, the hydrosilylation reaction probably occurs via true homogeneous catalytic intermediates. Many such systems, including one based on Wilkinson's catalyst, have been reported. An interesting aspect of such reactions is that hydrosilylation of —C=O and —C≡N bonds could be effectively carried out. Such transformations may have potential applications in the syntheses of fine chemicals and intermediates.

7.8.1   Catalytic Cycle and Mechanism

The first mechanistic proposal for the hydrosilylation reaction where mononuclear homogeneous catalytic intermediates are assumed is known as the Chalk–Harrod mechanism. The catalytic cycle in a slightly modified form is shown in Fig. 7.16. All steps of this catalytic cycle belong to organometallic reaction types that we have encountered many times before. Thus conversions of 7.60 to 7.61, 7.61 to 7.62, and 7.62 to 7.59 are examples of oxidative addition of $HSiR_3$, insertion of alkene into an M–Si bond, and reductive elimination, respectively.

As already mentioned, the highly active hydrosilylation catalysts derived from H_2PtCl_6 has been shown to be colloidal in nature. Colloid formation from the precatalyst $(COD)PtCl_2$ takes place according to reaction 7.28. This reaction obviously is not a stoichiometric one. The products cyclo-octene and cyclo-octane are formed by hydrogenation of 1,5-COD catalyzed by the platinum colloid. The hydrogen generated in the reaction comes from the silane reagent.

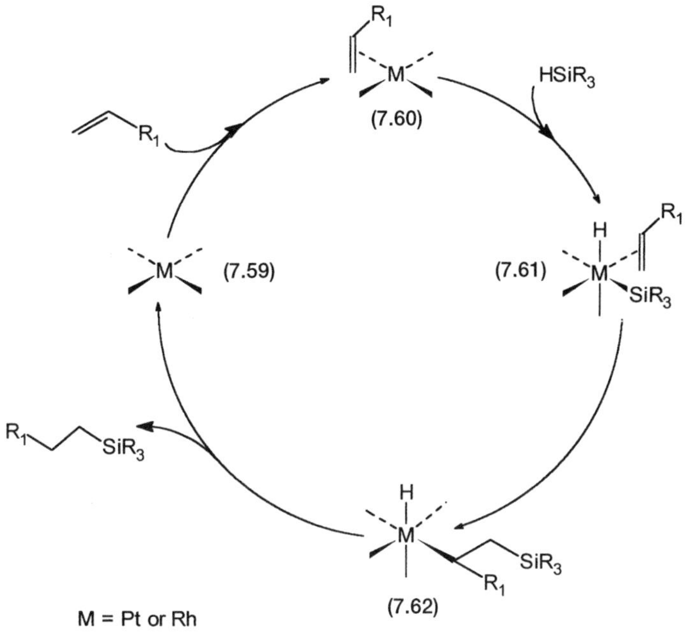

Figure 7.16 Proposed hydrosilylation mechanism with soluble catalysts of platinum or rhodium.

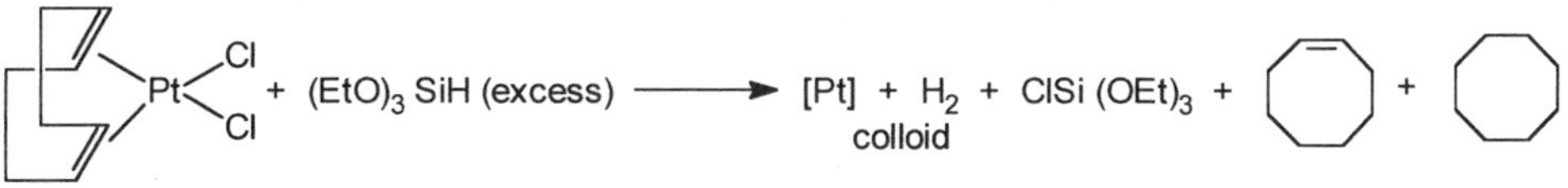

$$(7.28)$$

In recent years a variety of spectroscopic and other techniques have been employed to investigate and monitor hydrosilylation reactions. The techniques include multinuclear NMR, transmission electron microscopy, extended X-ray absorption fine structure (EXAFS), etc. Results from these experiments indicate that depending on the precatalyst, colloids and/or mononuclear complexes take part as catalytic intermediates.

7.9   C–C COUPLING AND CYCLOPROPANATION REACTIONS

Soluble palladium complexes catalyze a large number of reactions of the type 7.29 and 7.30. These reactions are often referred to as Heck reactions. The precatalyst is usually PdL_4 (L = PPh_3) or a combination of $Pd(OAc)_2$ with L. The base B could be organic or inorganic.

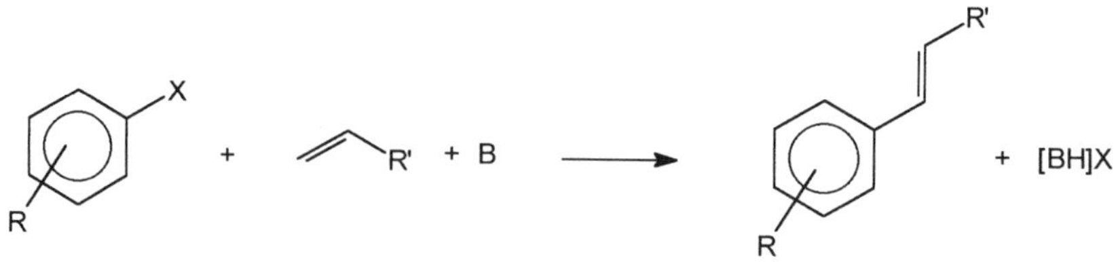

(7.29)

(7.30)

Although these reactions are highly versatile, so far they have not found significant industrial applications. The main reasons for this are relatively low thermal stabilities and turnover numbers of the catalytic systems, and the salt waste problem. However, Hoechst is trying to develop an industrial process for the manufacture of 6-methoxy 2-vinylnaphthalene by reaction 7.31. The precatalyst used is the dimeric phosphapallada cycle 7.63, which operates below a pressure of 20 atm and temperature of 130°C.

(7.31)

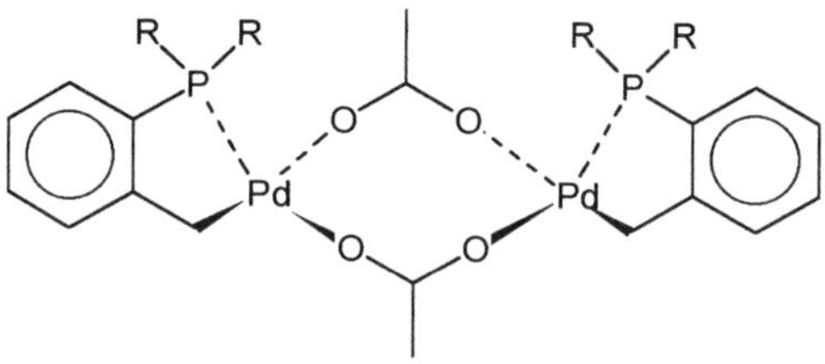

(R = o-tolyl)

(7.63)

Another variant of this type reaction has recently been utilized by Novartis to manufacture an intermediate for Prosulfuron® (see Table 1.1). The synthetic scheme is shown by reaction 7.32. The soluble palladium-complex-catalyzed C–C coupling between a diazonium salt and a fluoro alkene gives the required intermediate. This is then converted to the final product by standard reactions of synthetic organic chemistry.

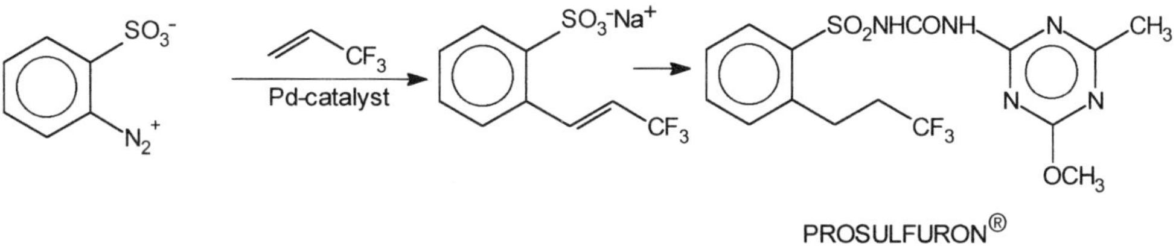

PROSULFURON®

(7.32)

Metal-catalyzed cyclopropanation of an alkene by a diazo compound, reaction 7.33, is another reaction where new C–C bonds are formed. This reaction finds use in the industrial manufacture of synthetic pyrethroids. The precatalysts for carbene addition reactions are coordination complexes of copper or rhodium. It should be noted that reaction 7.33 gives a mixture of isomers (*syn* plus *anti*) of the cyclopropane derivative. However, with some chiral catalysts, only one optical isomer with good enantioselectivity is obtained (see Section 9.5).

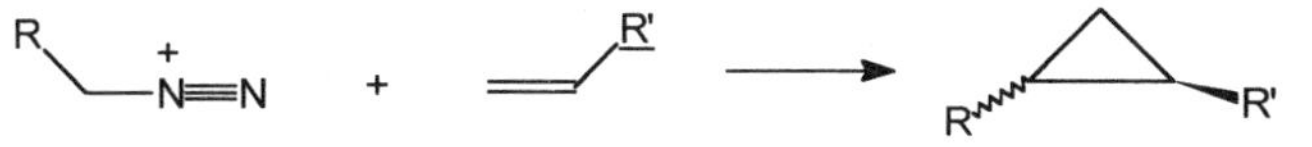

(7.33)

7.9.1 Catalytic Cycle for the Heck Reaction

A general catalytic cycle proposed for Heck reaction is shown in Fig. 7.17. While all the steps in the catalytic cycle have precedents, the proposed reaction mechanism lacks direct evidence. The basic assumption is that under the reaction conditions, the precatalyst is converted to 7.64, a coordinatively unsaturated species with palladium in the zero oxidation state. Oxidative addition of ArX, followed by alkene coordination, leads to the formation of 7.65 and 7.66, respectively. Alkene insertion into the Pd–C bond followed by β-H abstraction gives 7.67 and 7.68, respectively. Reductive elimination of HX, facilitated by the presence of base B, regenerates 7.64 and completes the catalytic cycle. The C–C coupled product is formed in the 7.67 to 7.68 conversion step.

There are two main uncertainties associated with this general mechanism. First, there are a number of C–C coupling reactions where there is no direct evidence for the reduction of the Pd(II) precatalyst into a zero-valent palladium species. Second, like the hydrosilylation system, a number of these reactions may involve colloidal palladium. Also, the general catalytic cycle needs to be substantially modified to rationalize the successful use of 7.63 as a precatalyst.

7.9.2 Catalytic Cycle for Cyclopropanation

Two mechanisms have been considered for metal-assisted cyclopropanation of alkenes by diazo compounds. These are shown by reactions 7.34 and 7.35.

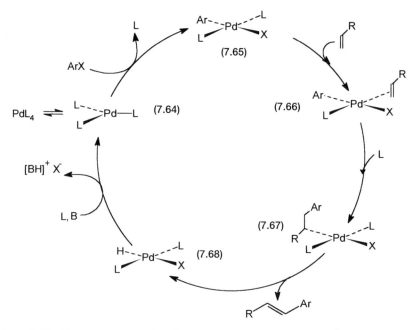

Figure 7.17 Proposed mechanism for C–C coupling (Heck reaction). Direct evidence for the catalytic steps are lacking.

Reaction 7.34 involves a metal–carbene intermediate, while reaction 7.35 involves nucleophilic attack by the diazo compound to the coordinated alkene. With a rhodium–porphyrin catalyst direct spectroscopic evidence has been obtained for the carbene pathway (see Section 2.5.2).

$$'M' \xrightarrow[- N_2]{+ RCHN_2} M{=}\overset{}{\underset{R}{\diagup}} \xrightarrow{} \quad 'M' + \underset{R \quad R'}{\triangle} \quad (7.34)$$

$$'M' \xrightarrow{} \quad M{-}\| \xrightarrow[- N_2]{+ RCHN_2} \quad 'M' + \underset{R \quad R'}{\triangle} \quad (7.35)$$

The catalytic cycle proposed for the rhodium–porphyrin-based catalyst is shown in Fig. 7.18. In the presence of alkene the rhodium–porphyrin pre-catalyst is converted to 7.69. Formations of 7.70 and 7.71 are inferred on the basis of NMR and other spectroscopic data. Reaction of alkene with 7.71 gives the cyclopropanated product and regenerates 7.69. As in metathesis reactions, the last step probably involves a metallocyclobutane intermediate that collapses to give the cyclopropane ring and free rhodium–porphyrin complex. This is assumed to be the case for all metal-catalyzed diazo compound-based cyclopropanation reactions.

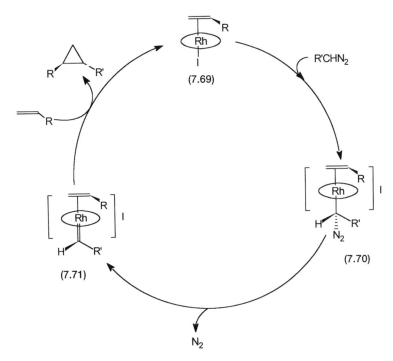

Figure 7.18 Catalytic cycle for the cyclopropanation of alkene by diazo compounds with a rhodium–porphyrin complex as the precatalyst.

PROBLEMS

1. In Figs. 7.1 and 7.2 identify oxidative addition, insertion, and reductive elimination steps. Show the formation of 7.7 from 7.6 by an electron pair (curly arrow) pushing formalism.

Ans. Fig. 7.1: no OA or RE, 7.2 to 7.3 insertion of alkene into M–H; Fig. 7.2: 7.6 to 7.7 OA, 7.7 to 7.5, RE. Reaction 7.1.

2. "The mechanism in Fig. 7.3 is applicable without modification for the hydrogenation of c-hexene, α-acetamidocinamic acid, and styrene by Wilkinson's catalyst." Discuss the validity of this statement.

Ans. A number of the mechanistic steps are common, but the sequence of reactions and observable intermediates are different.

3. Do oxidative addition (OA) and reductive elimination (RE) reactions have to be present in the catalytic cycle for all metal-catalyzed hydrogenation reactions?

Ans. No. Other mechanisms (radical, heterolytic, etc.) may be available with transition metal complexes. With lanthanide complexes in the highly stable 3+ oxidation state, an OA/RE-based mechanism is not possible.

4. From the rate expression given by 7.3 how could *a* be calculated graphically? How would the rate change if a very large amount of L was added externally?

Ans. A plot of rate against [Rh] at constant [H$_2$], [S], and externally added [L] would give a straight line of zero intercept. Such straight lines corresponding to different amounts of externally added [L] will have different slopes. A plot of the reciprocal of these slopes versus [L] will give a straight line with intercept = *a*. With a large amount of L the rate will be suppressed; at [L] = ∞ , rate = 0.

5. Draw a detailed catalytic cycle for the isomerization of 3-methyl-1-pentene by HRh(CO)L$_3$. How many isomers are expected and with what selectivity?

Ans. Two more isomers. Less of 2-ethyl-1-butene due to steric constraints at the Rh center.

6. A mixture of 1-butene and perdeuterated 1-butene are left over a solution of NiL$_3$ (L = PPh$_3$) for a long time. What are the expected products?

Ans. Formation of a randomly scrambled mixture of deuterated 1- and 2-butene by the allyl route.

7. The reaction between PPh$_3$ and PhCOCH$_2$Br followed by treatment with NaH gives an organic reaction product, which is then reacted with molar quantities of Ni(COD)$_2$ and PPh$_3$. What is the expected product, and what is the mechanism of its formation?

Ans. 7.17. Formation of Wittig reagent, followed by oxidative addition on Ni°, giving Ni–Ph and Ni–O bonds.

8. Starting from 1,5-COD and ethylene and using homogeneous and heterogeneous catalytic steps, how could *n*-C$_{10}$H$_{21}$CHO (a perfume additive) be made?

Ans. 1,5-COD to cyclooctene; cyclooctene plus ethylene metathesis to give C$_{10}$ dialkene, which on selective hydrogenation followed by hydroformylation will give the aldehyde.

9. In Fig. 7.7, β-elimination from two intermediates results in the formation of four olefins. Identify them.

Ans. 7.22(A), 7.23(A).

10. Write reactions of the type 7.8 and 7.9 to show the formation of 7.26, 7.27. The complex 7.30 is formed in the dimerization of an alkene. What is the alkene, and what is its polymer?

Ans. Isoprene, rubber.

11. In butadiene di- and trimerization with nickel catalyst and (*o*-PhC$_6$H$_4$O)$_3$P as a stabilizing ligand, the yields of cyclo-octadiene and cyclo-dodecatriene are 96% and 0.2%. However, if $(p\text{-}PhC_6H_4O)_3P$ is used as the ligand, the yields are 65% and 18%. Explain this observation.

Ans. The more bulky ortho-substituted phosphite favors sterically less demanding dimerization. The para-substituted phosphite is less discriminating.

12. Hercules sells a polymer made from dicyclopentadiene by ROMP. Draw its structure.

Ans. The "norborene"-like double bond takes part in metathesis.

13. Draw a catalytic cycle to show the disproportionation of $2R_1CH\!=\!CHR_2$ to $R_1CH\!=\!CHR_1$, and $R_2CH\!=\!CHR_2$.

Ans. Same as Fig. 7.11 with the directions of all arrows reversed.

14. In the isomerization of 2M3BN to 3PN, 7.46 and 7.47 are supposed to be involved; that is, formation of free butadiene is proposed. Starting with perdeuterated butadiene and HCN, based on this mechanism what would be the expected pattern of isotopic incorporation?

Ans. Incorporation of H in the 1,4-positions of butadiene.

15. In the hydrocyanation reaction, what are the products of Markovnikov and anti-Markovnikov additions? Although 2PN is thermodynamically about 50 times more stable than 4PN, how in the isomerization of 3PN to 4PN is its formation avoided?

Ans. Markovnikov: 2M3BN, MGN, anti-Markovnikov: 3PN, ADN. Rate of isomerization of 3PN to 4PN is faster than that of 3PN to 2PN.

16. Show how the final product of 7.14 could react with styrene to regenerate the original carbene species. What would be the organic product of this reaction?

Ans. Formation of a metallocyclobutane where the C attached to Ph is attached to Ru. Two vinyl groups in the 2 and 3 positions with respect to R containing carbon (see J. A. Tallarico et al., *J. Am. Chem. Soc.* **119**, 7157–58, 1997).

17. Sch 38516 is a natural product with antifungal properties. A synthetic scheme for this compound involves ring-closure metathesis to generate a 14-membered ring. How many carbon atoms are there in the precursor, and what is the other product?

Ans. 16, ethylene (see Z. Xu et al., *J. Am. Chem. Soc.* **119**, 10302–16, 1997).

18. In reactions 7.25 and 7.27 identify selectivity if any, and its origin.

Ans. The C–Si bond is formed with the less sterically hindered C atom.

19. Using one of the reactions discussed in this chapter, how could poly(1,4-phenylene vinylene), i.e., $[CH\!=\!CH\!-\!C_6H_4\!-\!CH\!=\!CH]_n$, be made?

Ans. Heck reaction with 1,4-dibromobenzene and ethylene.

20. The synthesis of Prosulfuron®, reaction 7.32 involves hydrogenation of the intermediate with Pd/C catalyst followed by other reactions. Suggest a method that would address both the problems of separation of the homogeneous catalyst and the subsequent hydrogenation step.

Ans. In situ generation of Pd–C from the soluble catalyst (see P. Baumeister et al., *Chimia* **51**, 144–46, 1997).

21. The complex RuCl₂L₃ (L = PPh₃) is sequentially treated with molar quantities of (2-isopropoxy phenyl)diazomethane and Pcy₃ (cy = cyclohexyl). What is the reaction product, and what reaction may it catalyze?

Ans. RuCl₂(=CHR)(Pcy₃), where R = 2-isopropoxy phenyl. The oxygen atom coordinates to Ru. Alkene metathesis (see J. S. Kingsbury et al., *J. Am. Chem. Soc.* **121**, 791–799, 1999).

BIBLIOGRAPHY

For all the sections
Books

Most of the books listed under Sections 1.3 and 2.1–2.3.4 contain information on hydrogenation, isomerization, etc. and should be consulted. Specially useful are the book by Parshall and Ittel and Sections 2.2, 2.5, and 2.6 of Vol. 1 of *Applied Homogeneous Catalysis with Organometallic Compounds*, ed. by B. Cornils and W. A. Herrmann, VCH, Weinheim, New York, 1996.

For nickel-based reactions (Sections 7.4, 7.5, 7.7) also see *The Organic Chemistry of Nickel*, P. W. Jolly and G. Wilke, Academic Press, New York, 1974.

Articles

Sections 7.1 to 7.3

B. R. James, in *Advances in Organometallic Chemistry*, ed. by F. G. A. Stone and R. West, Academic Press, Vol. 17, 1979, pp. 319–406.

F. H. Jardine, *Progr. Inorg. Chem.* **28**, 63–202, 1981.

B. R. James, in *Comprehensive Organometallic Chemistry*, ed. by G. Wilkinson, F. G. A. Stone, E. W. Abel, Pergamon Press, Vol. 8, 1982, pp. 285–369.

C. A. Tolman and J. W. Faller, in *Homogeneous Catalysis with Metal Phosphine Complexes*, ed. by L. H. Pignolet, Plenum Press, New York, 1983, pp. 13–109.

For NMR studies, see J. M. Brown et al., *J. Chem. Soc. Perkin Trans. II*, 1589–96, 1597–607, 1987.

For ab initio calculations, see C. Daniel et al., *J. Am. Chem. Soc.* 3473–3787, 1988.

Section 7.4

W. Keim et al., in *Comprehensive Organometallic Chemistry*, ed. by G. Wilkinson, F. G. A. Stone, E. W. Abel, Pergamon Press, Vol. 8, 1982, pp. 371–462.

D. Vogt, in *Applied Homogeneous Catalysis with Organometallic Compounds* (see under Books), Vol. 1, pp. 245–57.

E. F. Lutz, *J. Chem. Edu.* **63**, 202–3, 1986.

P. W. Jolly, in *Comprehensive Organometallic Chemistry*, ed. by G. Wilkinson, F. G. A. Stone, E. W. Abel, Pergamon Press, Vol. 8, 1982, pp. 615–48.

W. Keim, *Angew. Chem. Int. Ed.* **29**, 235–44, 1990; U. Muller et al., *ibid.* **28**, 1011–13, 1989.

Section 7.5

Y. Chauvin and H. Olivier, in *Applied Homogenous Catalysis with Organometallic Compounds* (see under Books), Vol. 1, pp. 258–68; G. Wilke and A. Eckerle, *ibid.*, 358–73.

B. Bogdanovic, in *Advances in Organometallic Chemistry*, ed. by F. G. A. Stone and R. West, Academic Press, Vol. 17, 1979, pp. 105–40.

A. C. L. Su, in *Advances in Organometallic Chemistry*, ed. by F. G. A. Stone and R. West, Academic Press, Vol. 17, 1979, pp. 269–318.

Section 7.6

R. H. Grubbs, in *Comprehensive Organometallic Chemistry*, ed. by G. Wilkinson, F. G. A. Stone, E. W. Abel, Pergamon Press, Vol. 8, 1982, pp. 499–552.

J. S. Moore, in *Comprehensive Organometallic Chemistry*, ed. by E. W. Abel, F. G. A. Stone, G. Wilkinson, Pergamon Press, Vol. 12, 1995, pp. 1233–300.

T. J. Katz, in *Advances in Organometallic Chemistry*, ed. by F. G. A. Stone and R. West, Academic Press, Vol. 16, 1977, pp. 283–317.

N. Calderon et al., in *Advances in Organometallic Chemistry*, Vol. 17, 1979, pp. 449–91.

J. C. Mol, in *Applied Homogeneous Catalysis with Organometallic Compounds* (see under Books), Vol. 1, pp. 318–32.

R. H. Grubbs et al., *Acc. Chem. Res.* **28**, 446–52, 1995.

The references in answers to Problem 16, 17, and 21.

Section 7.7

C. A. Tolman et al., in *Advances in Catalysis*, ed. by D. D. Eley, H. Pines, and P. B. Weisz, Academic Press, New York, Vol. 33, 1985, pp. 2–47.

C. A. Tolman, in *J. Chem. Edu.* **63**, 199–201, 1986.

Section 7.8

J. L. Speier, in *Advances in Organometallic Chemistry*, ed. by F. G. A. Stone and R. West, Academic Press, Vol. 17, 1979, pp. 407–48.

J. F. Harrod and A. J. Chalk, in *Organic Synthesis via Metal Carbonyls*, ed. by I. Wender and P. Pino, Wiley, New York, 1977, p. 163.

L. N. Lewis, *J. Am. Chem. Soc.* **112**, 5998–6004, 1990; J. Stein et al., *J. Am. Chem. Soc.* **121**, 3693–703, 1999.

Section 7.9

W. A. Herrmann, in *Applied Homogeneous Catalysis with Organometallic Compounds* (see under Books), Vol. 1, pp. 712–32.

A. F. Noels and A. Demonceau, *ibid.*, pp. 733–46.

For mechanistic studies with rhodium porphyrin, see J. L. Maxwell et al., *Science* **256**, 1544–47, 1992.

CHAPTER 8

OXIDATION

8.1 INTRODUCTION

Homogeneous catalysts are used for many large-scale oxidation processes. Some of the most important large-scale oxidation processes are acetaldehyde from ethylene, adipic acid from cyclohexane, terephthalic acid from p-xylene, and propylene oxide from propylene. The mechanisms of these reactions are very different and can be broadly classified into three categories. The first reaction, conversion of ethylene to acetaldehyde, involves organometallic and redox chemistry of palladium. The nucleophilic attack by water on coordinated ethylene being a central reaction of the overall process. Oxidation of cyclohexane and p-xylene by air, on the other hand, are chain reactions of organic radicals. In these reactions soluble cobalt and manganese ions catalyze the initiation steps. Reactions such as these, where the organic substrates are directly oxidized by air or dioxygen, are often called *autoxidation reactions*. The last reaction, conversion of propylene to propylene oxide, involves selective oxygen atom transfer chemistry, which has undergone a remarkable growth in the past two decades. The source of the oxygen atom in this type of reaction is not dioxygen but some other oxidizing agent, such as an organic hydroperoxide.

Apart from these large-scale oxidation reactions, there are many other reactions where catalytic oxidation is preferred either due to environmental considerations or high selectivity. An example of how environmental considerations come into play is shown in Fig. 8.1. The old route from benzene to hydroquinone by (a) generates salts such as $MnSO_4$, $FeCl_2$, Na_2SO_4, $NaCl$, etc., whose total weight is about ten times that of the product. In contrast, the catalytic route as shown by (b) generates considerably less salt than the product.

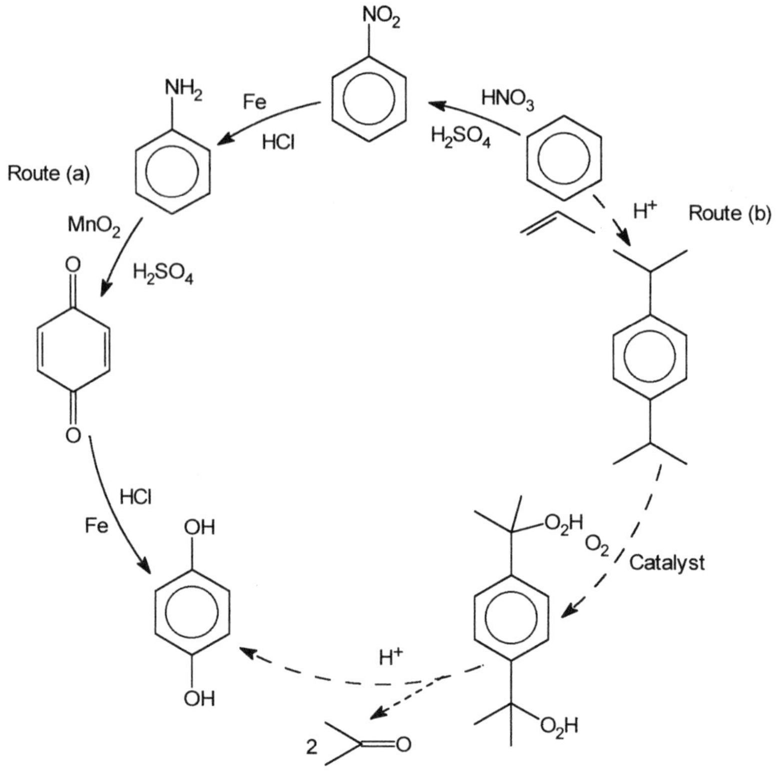

Figure 8.1   The old and the new manufacturing processes for *p*-dihydroxy benzene. Route (a) is the old process, which involves stoichiometric oxidation, while route (b) involves catalytic oxidation. The amount of solid waste generated in (a) is an environmental hazard.

Many pharmaceutical and fine chemical intermediates have functional groups that could be easily derived from epoxides. In recent years, applications of homogeneous catalysis have been vigorously explored for the selective synthesis of epoxides.

8.2   WACKER OXIDATION

Conversion of ethylene to acetaldehyde with a soluble palladium complex was one of the early applications of homogeneous catalysis. Traditionally, acetaldehyde was manufactured either by the hydration of acetylene or by the oxidation of ethanol. As most of the acetic acid manufacturing processes were based on acetaldehyde oxidation, the easy conversion of ethylene to acetaldehyde by the Wacker process was historically a significant discovery. With the advent of the methanol carbonylation process for the manufacture of acetic acid, the industrial importance of the Wacker process has diminished.

However, the Wacker process still remains an important homogeneous catalytic reaction for several reasons. First of all, like some of the other processes discussed earlier, Wacker process is a good example of some of the fundamental concepts and reactions of organometallic chemistry. Second, though vinyl acetate is industrially manufactured by a heterogeneous catalytic route, the underlying chemistry is closely related to that of the Wacker process. Third, it is likely that using similar chemistry, an industrial process for the conversion of butenes to methyl ethyl ketone will be developed in the near future. Finally, the scale of manufacture of acetaldehyde from ethylene, by this homogeneous catalytic process is industrially still significant.

8.2.1 The Background Chemistry

The Wacker process is based on three reactions: oxidation of ethylene by Pd^{2+} in water, oxidation of Pd^0 to Pd^{2+} by Cu^{2+}, and oxidation of Cu^+ by dioxygen to Cu^{2+}. These reactions are shown by 8.1 to 8.3. Note that if the three reactions are summed up, the net reaction becomes one mole of ethylene and half a mole of oxygen, giving one mole of acetaldehyde. Both palladium and copper ions shuttle between two oxidation states to act as the catalysts.

$$Pd^{2+} + H_2O + \; \| \; \longrightarrow \; \overset{}{\underset{H}{>}}\!\!=\!O \; + \; Pd^0 \; + \; 2H^+ \qquad (8.1)$$

$$Pd^0 + 2Cu^{2+} \; \longrightarrow \; Pd^{2+} + 2Cu^+ \qquad (8.2)$$

$$2Cu^+ + 2H^+ + \tfrac{1}{2}O_2 \; \longrightarrow \; 2Cu^{2+} + H_2O \qquad (8.3)$$

Similar net reaction can be effected between ethylene and acetic acid to give vinyl acetate. This is shown by 8.4. Instead of acetic acid, if the reaction is carried out in alcohols, then the products are vinyl ethers. Oxidation of internal alkenes leads to the formation of ketones. Reactions 8.5 and 8.6 show these conversions. It is reasonable to assume that in all these cases the basic mechanism is similar to that of ethylene oxidation to acetaldehyde.

$$(8.4)$$

$$ROH + \| + \tfrac{1}{2}O_2 \; \longrightarrow \; \diagup\!\!\!\diagup OR \; + \; H_2O \qquad (8.5)$$

$$(8.6)$$

The reaction media for Wacker-type reactions are highly corrosive. This is due to the presence of free acids (acetic acid for vinyl acetate), ions like Cl⁻, and dioxygen. For any successful technology development, the material of construction for the reactors is a major point of concern (see Section 3.1.4). Some progress in this respect has recently been made by the incorporation of heteropolyions such as $[PV_{14}O_{42}]^{9-}$ in the catalytic system. The heteropolyions probably act as redox catalysts. A seminonaqueous system is used for this modified catalytic system, and the use of low pH for dissolving copper and palladium salts is avoided.

8.2.2   Catalytic Cycle and Mechanism

The catalytic cycle proposed for ethylene to acetaldehyde is shown in Fig. 8.2. The tetrachloro palladium anion 8.1 is used as the precatalyst. Conversion of 8.1 to 8.3 involves substitution of two chloride ligands by ethylene and water. Nucleophilic attack on coordinated ethylene leads to the formation of 8.4. The latter then undergoes substitution of another Cl⁻ ligand. Conversion of 8.5 to 8.6 involves β-hydride abstraction and coordination by vinyl alcohol. Intramolecular hydride attack to the coordinated vinyl group leads to the formation of 8.7. The latter eliminates acetaldehyde, proton, and Cl⁻ and in the process is reduced to a palladium complex of zero oxidation state.

The sum of all these steps is the net reaction 8.1. To make the reaction catalytic, 8.8 must be converted back to 8.1. This is achieved either in the same reactor or in another one by oxidizing zero-valent palladium with copper(II) chloride (Eq. 8.2) and the resultant cuprous chloride by dioxygen (Eq. 8.3).

The mechanistic evidence for the proposed catalytic cycle comes from kinetic and isotope-labeling studies. First, the rate law is as shown in Eq. 8.7. This indicates a preequilibrium that involves the displacement of two Cl⁻ and one H⁺ ions, that is, conversion of 8.1 to 8.4. It is generally agreed that conversion of 8.4 to 8.5 is the rate-determining step, which is consistent with the observed rate law. The conversion of 8.3 to 8.4 has been a matter of some controversy.

$$-\frac{d[C_2H_4]}{dt} = k\,\frac{[Pd^{2+}][C_2H_4]}{[H^+][Cl^-]^2} \tag{8.7}$$

In principle both *inter*molecular reactions between 8.3 and external water, and *intra*molecular reactions between coordinated ethylene and coordinated water could lead to the formation of 8.4. On the basis of experiments with deuterium-labeled ethylene followed by stereochemical analysis of the product, it appears that the *inter*molecular reaction with solvent water molecule is the pathway that is followed. This is shown in Fig. 8.3. Note, however, that the D-labeling results are valid only for a specific set of conditions. These conditions are not the same as in the industrial process. Intramolecular attack by

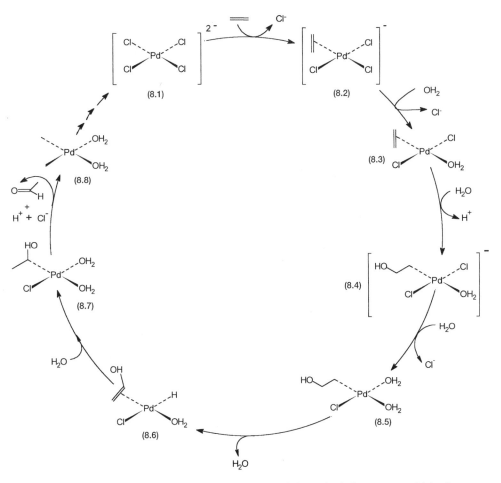

Figure 8.2 Catalytic cycle for the Wacker oxidation of ethylene to acetaldehyde.

coordinated water (or hydroxide) should take place in a *cis* manner but this is not observed.

Another important piece of mechanistic evidence comes from the fact that acetaldehyde obtained from ethylene and deuterium-labeled water does not have any deuterium incorporation. In other words, all four hydrogen atoms of the ethylene molecule are retained. This means vinyl alcohol, which would certainly exchange the hydroxyl proton with deuterium, cannot be a free short-lived intermediate that tautomerizes to acetaldehyde. However, in spite of many elegant synthetic accomplishments in organometallic chemistry, realistic model complexes of palladium for 8.4 to 8.7 remain unknown.

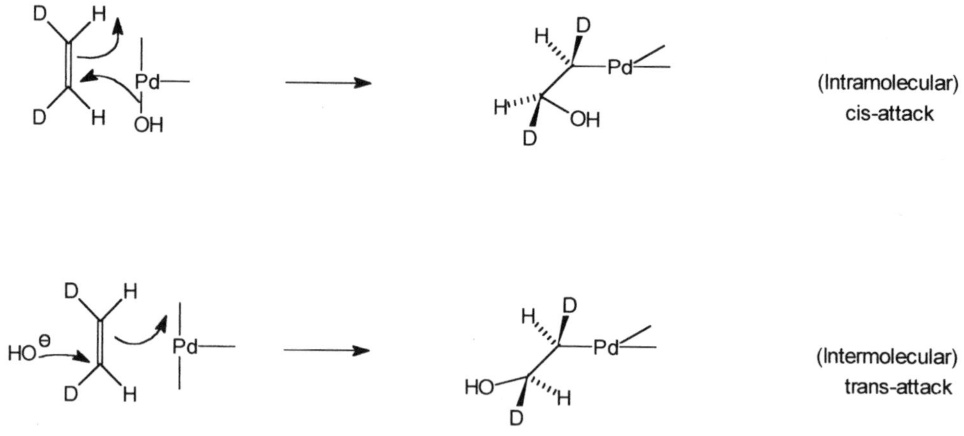

(Intramolecular)
cis-attack

(Intermolecular)
trans-attack

Figure 8.3 Stereochemistry of nucleophilic attack by HO⁻ on the coordinated ethylene. *Intra*molecular and *inter*molecular reactions should give *cis* and *trans* orientations of Pd–C and C–O bonds, respectively. D labeling helps to identify the stereochemistry of addition.

8.3 METAL-CATALYZED LIQUID-PHASE AUTOXIDATION

As mentioned earlier, soluble salts of cobalt and manganese catalyze oxidation of cyclohexane by oxygen to cyclohexanol and cyclohexanone. Cyclohexanol and cyclohexanone are oxidized by nitric acid to give adipic acid. The oxidation by nitric acid is carried out in the presence of V^{5+} and Cu^{2+} ions. These reactions are shown by Eq. 8.8. Adipic acid is used in the manufacture of nylon 6,6.

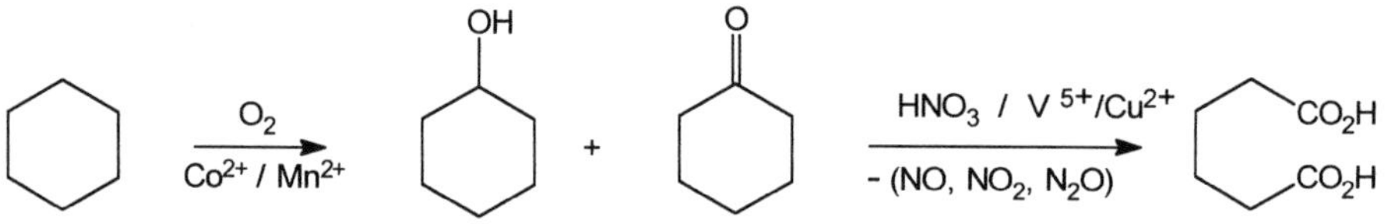

$$(8.8)$$

The oxidation of cyclohexanone by nitric acid leads to the generation of nitrogen dioxide, nitric oxide, and nitrous oxide. The first two gases can be recycled for the synthesis of nitric acid. Nitrous oxide, however, is an ozone depleter and cannot be recycled. Indiscriminate nitrous oxide emission from this process is therefore the cause of considerable concern. As shown by 8.9, part of the cyclohexanone can also be converted to the corresponding oxime and then to caprolactam—the monomer for nylon 6. Phthalic acids are one of the monomers for the manufacture of polyesters. As shown by Eq. 8.10, it is made by the oxidation of *p*-xylene. A general description of polyamides (nylons) and polyesters are given in Section 8.4.

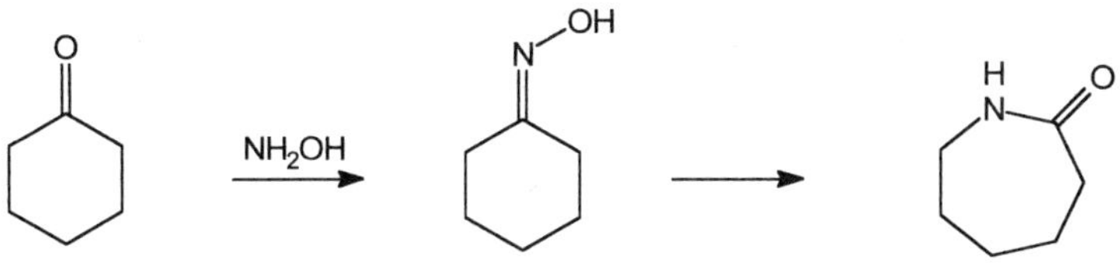

(8.9)

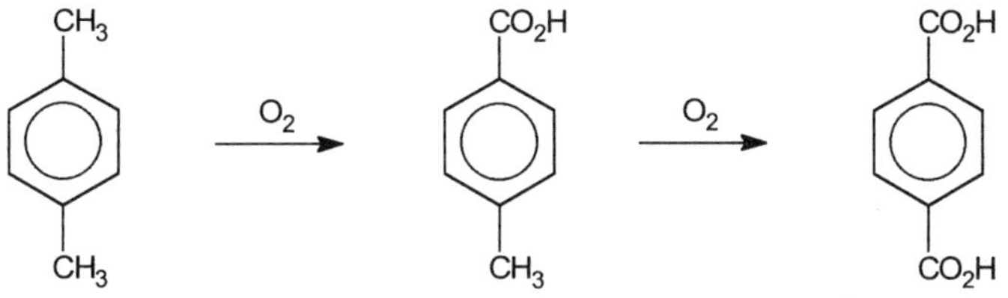

(8.10)

8.3.1   Mechanism of Autoxidation

Auto-oxidation processes consist of a very large number of simultaneous and consecutive radical reactions. Most of these reactions can be categorized into three basic types, and these are shown by 8.11 to 8.15. Note that for these reactions an organic radical initiator In_2 and no metal complex is used. The role of a metal ion in autoxidation processes is that of a catalyst for the initiation steps.

$$INITIATION: \quad In_2 \longrightarrow 2\, In\cdot \tag{8.11}$$

$$In^{\bullet} + RH \longrightarrow InH + R^{\bullet} \tag{8.12}$$

$$PROPAGATION: \quad R^{\bullet} + O_2 \longrightarrow RO_2^{\bullet} \tag{8.13}$$

$$RO_2^{\bullet} + RH \longrightarrow RO_2H + R^{\bullet} \tag{8.14}$$

$$TERMINATION: \quad 2RO_2^{\bullet} \longrightarrow Oxygenated\ Products \tag{8.15}$$

Organic hydroperoxides can be reduced (gain of electron) and oxidized (loss of electron) by metal ions that can easily shuttle between two oxidation states. This is shown by reactions 8.16 and 8.17. It is obvious that the combined effect of these two reactions is the metal-catalyzed decomposition of hydroperoxide to give water, alkoxo, and hydroperoxo radicals. Once these radicals are generated, oxidation can proceed according to the usual propagation and termination steps. In other words, if the metal ions can bring about reactions 8.16 and 8.17, reactions 8.13 to 8.15 would automatically follow. An important addition to the propagation steps under these conditions is shown by 8.19. Note that in In_2-initiated autoxidation $RO^{\bullet}$ is not formed in the initiation or propagation steps.

METAL CATALYZED
INITIATION:

$$RO_2H + M^{n+} \longrightarrow RO^{\bullet} + HO^- + M^{(n+1)+} \qquad (8.16)$$

$$RO_2H + M^{(n+1)+} \longrightarrow RO_2^{\bullet} + H^+ + M^{n+} \qquad (8.17)$$

$$RH + M^{(n+1)+} \longrightarrow R + H^+ + M^{n+} \qquad (8.18)$$

$$(M = Co, \; n+1 = 3)$$

ADDITIONAL PROPAGATION
STEP:

$$RO^{\bullet} + RH \longrightarrow ROH + R^{\bullet} \qquad (8.19)$$

The facility with which metal complexes bring about reactions 8.16 and 8.17 depends on several factors, one of the important ones being the half-cell potential (E°) of the $M^{n+}/M^{(n+1)+}$ couple. It should be remembered, however, that most E° values for metal ions have been measured in an aqueous environment. On complexation and in an organic liquid these values are expected to change substantially. The initial hydroperoxide required for metal-catalyzed decomposition, reactions 8.16 and 8.17, is normally present in trace quantities in most hydrocarbons.

With purified hydrocarbons where such trace impurities have been scrupulously removed, a long induction time precedes metal-catalyzed autoxidation. The mechanism of formation of initial RO_2H in the absence of a radical chain is not known in any detail, but metal ions do not seem to be involved. The most important point to note in this context is that various types of metal–dioxygen complexes have been isolated and fully characterized. However, such complexes do *not* seem to play any role in metal ion initiated autoxidation reactions.

With selected organic substrates direct oxidation by metal could be an additional initiation step. This is shown by reaction 8.18. The thermodynamic feasibility of such a reaction depends on the stability of the radical $R^{\bullet}$ as well as the E° value of the $M^{(n+1)+}/M^{n+}$ couple. The oxidation potential of Co^{3+}/Co^{2+} is as high as 1.82. As we will see, direct electron transfer from p-xylene to Co^{3+} is an important initiation step for the oxidation of p-xylene.

In a real system many reactions that fit into the general categories represented by 8.13 to 8.15 are possible. This is because the organic intermediates and products themselves may undergo further rearrangement, oxidation, and other reactions. Mechanistic studies for these reactions are therefore invariably based on kinetic models. In these models a set of reactions and associated rate constants are assumed. Through simulation and optimization methods the model is then refined so that best fit between observed and predicted data points are obtained.

The propagation step 8.14 is of special importance. Its rate depends on the R–H bond strength; the weaker the bond, the faster the reaction. This explains why autoxidation of aldehydes is much easier than that of saturated hydrocarbons. The C–H bond energies of –CHO– and –CH$_2$– groups are approximately 85 and 100 kcal mol^{-1}, respectively. This results in a large difference in the relative rates of the respective propagation steps.

8.3.2 Special Features of Cyclohexane Oxidation

The first step, oxidation of cyclohexane to cyclohexanol and cyclohexanone, follows the general mechanism outlined by reactions 8.13 to 8.17. Trace quantities of cyclohexyl hydroperoxide 8.9 can initiate the radical chain, where the radicals 8.10 and 8.11 take part in the propagation steps.

(8.9)   (8.10)   (8.11)

One of the termination steps that lead to the simultaneous formation of both cyclohexanol and cyclohexanone deserves attention. The peroxy radical dimerizes to give the intermediate 8.12. The dimer as shown by reaction 8.20, undergoes intramolecular rearrangement to give the products.

(8.12)   (8.20)

The oxidation of cyclohexane is accompanied by the oxidative degradation of the products. In other words, the products themselves undergo metal-initiated radical chain oxidation. For this reason oxidation is normally carried out at low conversions when the selectivity to the desired products is high. The catalysts used are a mixture of hydrocarbon soluble carboxylate salts of Co^{2+} and Mn^{2+} or Cr^{3+}. Due to the better solubility properties, salts of long-chain carboxylic acids such as 2-ethyl hexanoic or naphthenic acids are favored. As already mentioned, the primary role of the metal ions is to act as catalysts for the initiation steps. In other words, the metal ions undergo redox reactions with 8.9 to give 8.10 and 8.11.

The mixture of cyclohexanone and cyclohexanol is oxidized by nitric acid in the presence of V^{5+} and Cu^{2+} ions. This is shown in Fig. 8.4. Since some of the nitric acid is lost as N$_2$ and N$_2$O, the oxidation is "semistoichiometric" in nitric acid. However, the NO and NO$_2$ produced in this oxidation are recycled back.

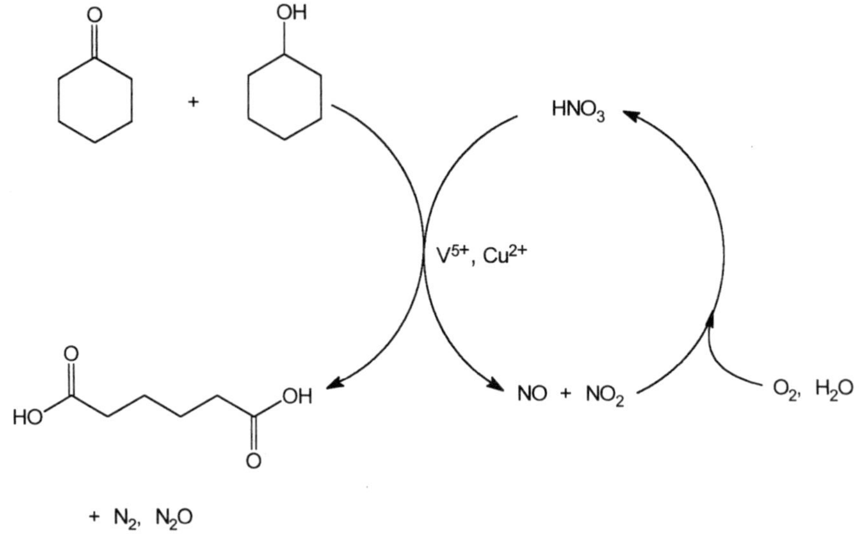

Figure 8.4   Oxidation of cyclohexanol and cyclohexanone by HNO_3 to give adipic acid. V^{5+} and Cu^{2+} are in catalytic amounts. NO and NO_2 are recycled, but some HNO_3 is lost as N_2 and N_2O (a greenhouse gas).

The roles of vanadium and copper are complex and not fully understood. A large number of intermediates, including the nitroso and the diketo compound as shown by reaction 8.21 are formed. The ion VO_2^+ stoichiometrically oxidizes the diketo compound to adipic acid. This is shown by reaction 8.22.

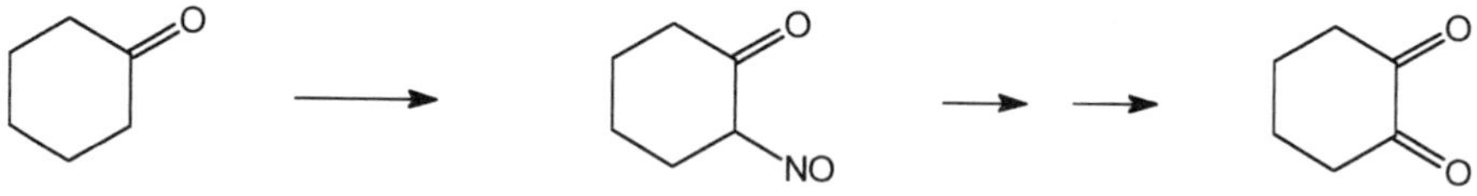

(8.21)

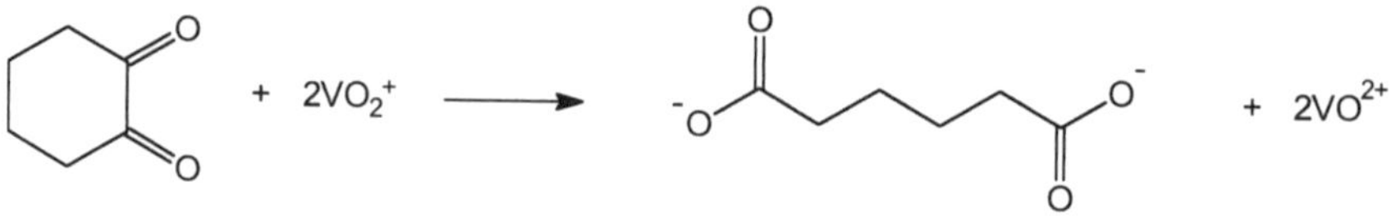

(8.22)

As shown by reaction 8.23, VO^{2+} species thus generated is converted back to VO_2^+ by nitric acid.

$$VO^{2+} + NO_3^- \rightarrow VO_2^+ + NO_2 \qquad (8.23)$$

The role of copper ion seems to be facilitating further nitration of the nitroso compound. As shown by reaction 8.24, the nitro, nitroso ketone gets converted to adipic acid. In the absence of copper other degradation pathways leading to unwanted products dominate.

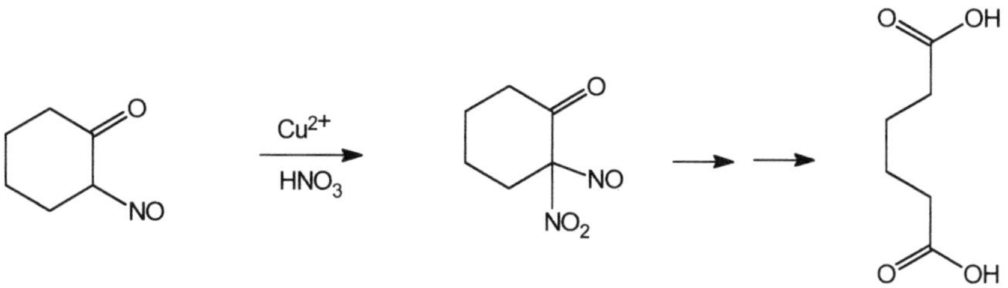

(8.24)

8.3.3 Special Features of *p*-Xylene Oxidation

The most important mechanistic difference between cyclohexane and xylene oxidation is that in the former there is no direct electron transfer from the substrate to the metal ion. In other words, in cyclohexane oxidation the initiation step 8.18 has no role to play. In contrast for all methylbenzene derivatives, xylenes included, this pathway is of importance for chain initiation. As shown by 8.25, formation of the radical species takes place through a radical cation.

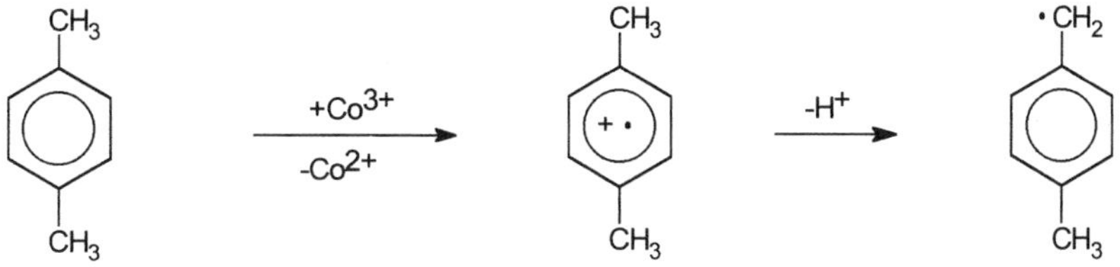

(8.25)

The initial products are the *p*-methyl benzyl alcohol and *p*-methyl benzaldehyde. Under the reaction conditions these are further oxidized, first to *p*-toluic acid, and ultimately to terephthalic acid. The alkoxy and peroxy radicals that take part in the oxidation of *p*-xylene to *p*-toluic acid are shown by 8.13 to 8.16.

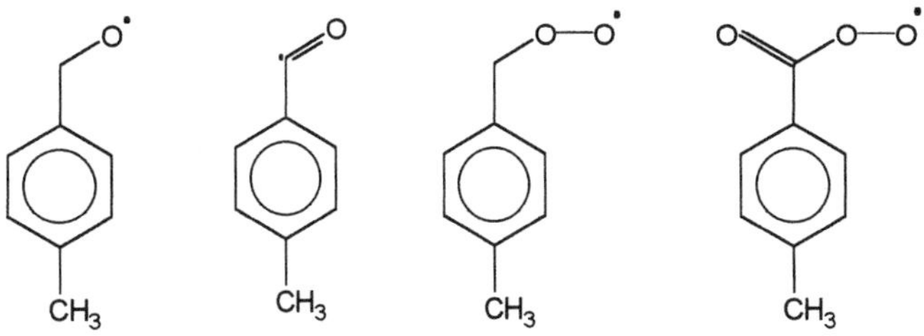

(8.13)   (8.14)   (8.15)   (8.16)

Commercially two main processes, that of Mid-century/Amoco and Dynamit Nobel/Hercules, are operated. In the former acetic acid is used as a solvent. Mixtures of cobalt and manganese bromide and acetate salts are used to catalyze the initiation step. The reaction conditions, a temperature of about 220°C and a pressure of 15 atm, are relatively severe. Under these conditions bromine and $^{•}CH_2CO_2H$ radicals are formed. These radicals can effect new initiation steps. In the overall process, though toluic acid is an intermediate, it is never isolated. The final isolated product is terephthalic acid (see reaction 8.10).

In the Dynamit Nobel/Hercules process no solvent is used. A mixture of cobalt and manganese ethyl hexanoate is used as the catalyst under relatively mild conditions, about 160°C and 7 atm pressure. The product under these conditions is toluic acid, which is isolated and then converted into the methyl ester. The important point to note is that under the operating conditions toluic acid does not undergo any further oxidation. This means that toluic acid is more difficult to oxidize than p-xylene. The methyl ester of toluic acid is then co-oxidized with p-xylene. The product obtained is monomethyl terephthalate, which by reaction with methanol is then converted to dimethyl terephthalate.

The advantage of the Amoco process is that high-purity terephthalic acid is produced in one step. The solubility of terephthalic acid in acetic acid is low; its separation with high purity by crystallization is therefore relatively easy. However, the corrosive nature of the acids and the relatively drastic conditions make it necessary to use special material of construction for the reactors.

8.4   POLYMERS (POLYESTERS AND POLYAMIDES) FROM AUTOXIDATION PRODUCTS

Poly(ethylene terephthalate) in short PET is a polyester. It is mainly used in the garment industry with or without natural cotton and has trade names such as Terylene®, Dacron®, etc. As the name indicates, it is a polymer between terephthalic acid (PT) and ethylene glycol. Both terephthalic acid and dimethyl terephthalate (DMT) can be used to make the polymer. A majority of the modern plants tend to use PT as the starting material because of the availability of high-purity PT on a large scale. Both PT and DMT are first converted to bis(hydroxy ethyl) terephthalate 8.17 (see reaction 8.26). For PT this is effected by a straightforward esterification reaction. For DMT a transesterification reaction catalyzed by zinc and manganese acetate is used.

(8.17)   (8.18)

(8.26)

The polycondensation reaction to give the desired polymer, 8.18, is carried out at high temperatures (>260°C) in the presence of Sb_2O_3 and Ti-alkoxide catalysts. The roles of the catalysts at the molecular level are not known in any detail. As shown by reaction 8.26, in the polycondensation reaction one molecule of ethylene glycol is also generated for each molecule of the monomer.

Nylon 6,6 and nylon 6 are polyamides. These polymers are used in carpets, in hosiery, and in certain cases as engineering plastics. Nylon 6,6 8.19, is the condensation product between adipic acid and 1,6-diamino hexane. Nylon 6 8.20 is made from caprolactam by ring-opening polymerization.

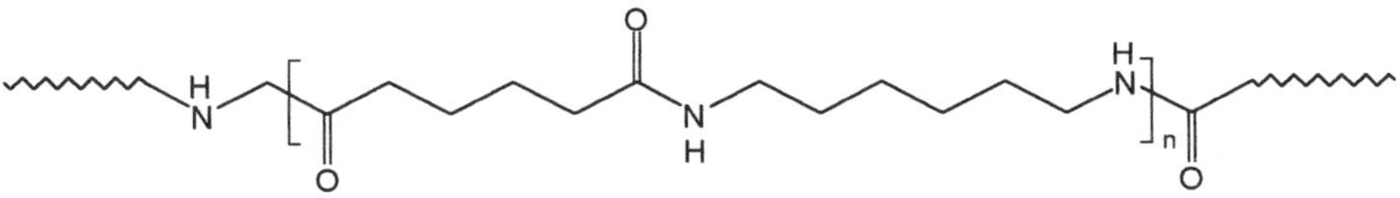

(8.19)

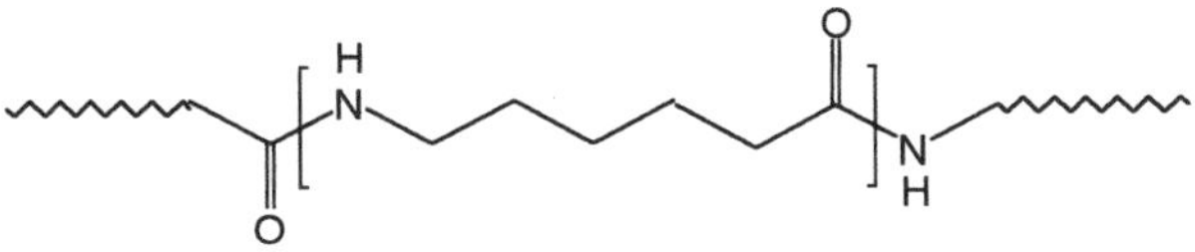

(8.20)

8.5 EPOXIDATION OF PROPYLENE

Propylene oxide is an important commmodity chemical. It is used in the manufacture of propylene glycol, glycerine, polyethers, etc. Ethylene oxide can be made selectively by the direct oxidation of ethylene over a heterogeneous catalyst, but propylene to propylene oxide by selective heterogeneous catalytic oxidation has so far not been achieved. As shown by 8.27, traditionally propylene oxide has been made by converting propylene to the chlorohydrin and then dehydrochlorinating it. The hydrochloric acid generated in the last step is neutralized with lime.

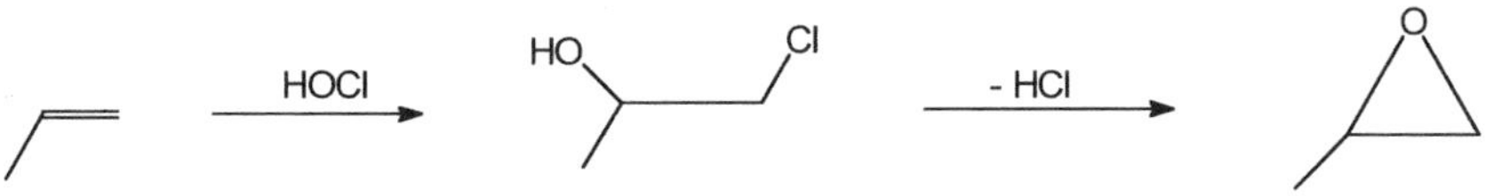

CHLOROHYDRIN

(8.27)

A homogeneous catalytic process, developed by Oxirane, uses a molybdenum catalyst that epoxidizes propylene by transferring an oxygen atom from tertiary butyl hydroperoxide. This is shown by 8.28. The hydroperoxide is obtained by the auto-oxidation of isobutane. The co-product of propylene oxide, *t*-butanol, finds use as an antiknock gasoline additive. It is also used in the synthesis of methyl *t*-butyl ether, another important gasoline additive. The overall superiority of the homogeneous catalytic process over the chlorohydrin route lies in better economics, as well as avoidance of generation of large quantities of chloride salt as waste (see Problem 5 in Chapter 1).

$$(8.28)$$

8.5.1 Catalytic Cycle and the Mechanism of Propylene Epoxidation

The proposed catalytic cycle is shown in Fig. 8.5. A variety of molybdenum compounds may be used as the precatalyst, and $Mo(CO)_6$ is shown as a representative one. Under the strong oxidizing conditions the precatalyst is oxidized to 8.21, a species that has molybdenum in a 6+ oxidation state and a cis-MoO_2^{2+} unit. The other ligands are two solvent molecules and hydroxo and/or alkoxo groups. In the absence of a solvent, positions occupied by S are occupied by t-butanol, the decomposition product of t-butyl hydroperoxide. The important points to note are that molybdenum is in its highest oxidation state (6+), and there are weakly bound solvent molecules.

On reaction with Bu^{t}OOH, 8.21 eliminates Bu^{t}OH and is converted to 8.22, which has a chelating t-butyl hydroperoxo ligand. Two mechanistic pathways from 8.22 have been proposed. In Path A propylene undergoes electrophilic attack by the distal oxygen atom, the oxygen atom away from the t-butyl group. As shown by reaction 8.29, Ti^{4+}, V^{5+}, and Mo^{6+} are known to be capable of effecting this type of oxygen atom transfer reaction. Note that coordination by propylene is not invoked if 8.23 is the proposed intermediate. The O atom transfer leads to 8.24, which has a weakly bound propylene oxide molecule. A solvent molecule displaces the coordinated propylene oxide. This regenerates 8.21, and the catalytic cycle is completed.

M^{n+} = Mo^{6+}, V^{5+}, Ti^{4+}

$$(8.29)$$

In the other mechanistic postulate, coordination by the alkene as in 8.25 followed by insertion into the Mo−O bond to give a metallocyclic species 8.26 is proposed. As shown by Path B, this metallocyclic species is then converted to 8.24 through an intramolecular rearrangement.

Species like 8.22 with V^{5+} as the metal ion has been isolated and characterized by X-ray studies (see Section 2.5.3). The reactivity of such complexes with alkenes has considerable similarities to the molybdenum-catalyzed epoxidation reaction. Kinetic studies with these model complexes indicate coordi-

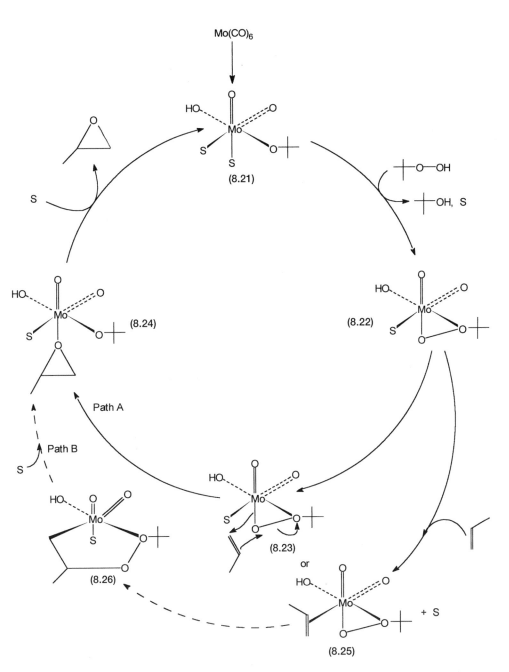

Figure 8.5 Catalytic cycle for the molybdenum-catalyzed oxidation of propylene by *t*-butyl hydroperoxide. Paths A and B are the two proposed alternative mechanisms for oxygen atom transfer. No evidence has been found for 8.26.

nation by the alkene as shown for 8.25. Metallocyclic complexes of Ti, V, or Mo of the type 8.26 have not been isolated, nor is there any spectroscopic evidence to suggest their involvement.

Unfortunately the macroscopic rate law cannot differentiate between Path A and Path B. The rate law in this and other related epoxidations is found to be

$$\text{Rate} = k[\text{catalyst}][\text{alkene}][\text{hydroperoxide}]$$

All the evidence taken together suggest possible coordination by propylene as in 8.25, but conversion of 8.25 directly to 8.24. In other words, there is no evidence for the involvement of a species such as 8.26. In asymmetric epoxidation of allylic alcohols, to be discussed in the next chapter, the alkene coordinates to the metal ion through the oxygen lone pair rather than the C–C double bond.

8.6   OXO COMPLEXES AS HOMOGENEOUS OXIDATION CATALYSTS

Metal complexes combined with a suitable oxygen atom donor can often act as versatile and selective oxidation catalysts. The oxygen atom donors commonly used are N-methyl morpholine oxide 8.27 (see reaction 8.30), organic hydroperoxides, sodium hypochlorite, hydrogen peroxide, etc. Three such reactions of industrial relevance are shown by 8.30, 8.31, and 8.32. Obviously these reactions are catalytic with respect to the metal complex but stoichiometric with respect to the oxygen atom donor.

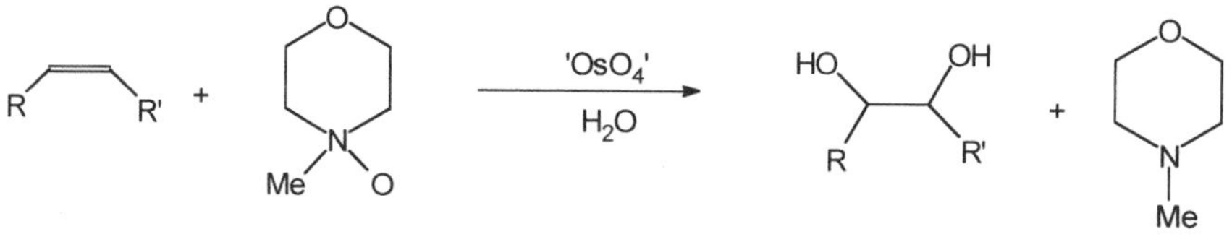

(8.27)

(8.30)

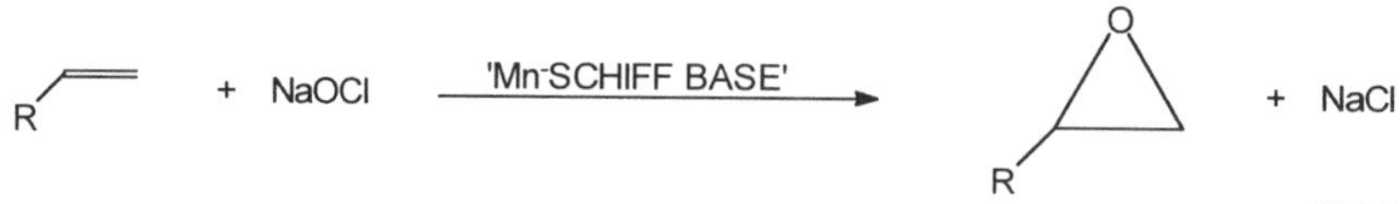

(8.31)

POLYPHENOLICS + H$_2$O$_2$ $\xrightarrow{\text{Mn·MACROCYCLE}}$ OXIDIZED POLYPHENOLICS + H$_2$O

(Tea Stain!)   (Bleached tea stain)

(8.32)

The first reaction, 8.30, is the classical alkene dihydroxylation with osmium tetroxide. In recent years the potential of this reaction has been vastly extended by making asymmetric catalytic dihydroxylation possible (see Section 9.4). The second reaction, 8.31, is an epoxidation reaction that does not involve organic hydroperoxide and high-valent metal ions such as Ti^{4+}, V^{5+}, or Mo^{6+}.

The importance of this reaction also lies in the fact that asymmetric epoxidation of alkenes other than allylic alcohols is possible with this catalytic system (see Section 9.3.4). The third reaction relates to catalysts developed by Unilever for improved detergent action in the presence of hydrogen peroxide. The important point to note is that catalytic intermediates with metal-oxo groups play a pivotal role in all these reactions.

Apart from potential industrial applications, homogeneous catalytic systems with metal–oxo intermediates have direct relevance to certain biological oxidation reactions. The biological reactions involve metalloenzymes such as Cyt P450, which has an iron–porphyrin complex in its active site. The enzyme catalyzes hydroxylation of a hydrocarbon by oxygen. This hydroxylation does not proceed through a radical-chain mechanism. There is sufficient evidence to indicate that a catalytic intermediate with an Fe=O group is responsible for the hydroxylation reaction.

8.6.1   Mechanism of Oxidation by Oxo Compounds

Mechanisms of reactions 8.30 and 8.31 are discussed in the next chapter. The mechanistic details of the Unilever reaction are unknown, but some interesting observations have been reported. The ligand 1,4,7-trimethyl-1,4,7-triazacyclononane 8.28 have been found to be particularly effective in forming manganese complexes that are active catalysts for reaction 8.32. One such complex is shown by 8.29.

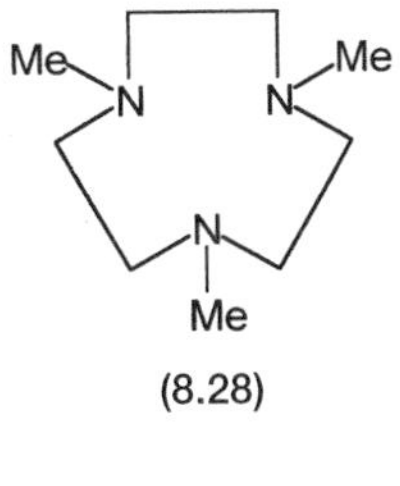

(8.28)

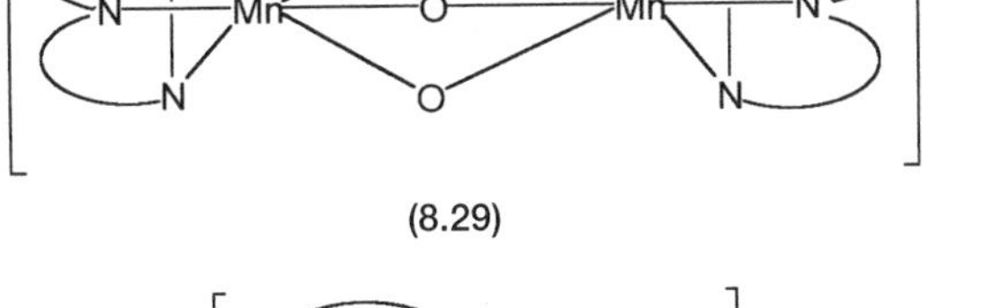

(8.29)

A tentative mechanism for the enhanced bleaching ability is shown by Fig. 8.6. Polyphenolics are largely responsible for tea stains. Complex 8.29 has both the manganese ions in 4+ oxidation states. When used in a detergent formulation, it oxidizes the polyphenolics and is reduced to a species containing one Mn^{4+} and one Mn^{3+} ion. This species is tentatively formulated as 8.30. The hydrogen peroxide generated in situ by the detergent probably oxidizes 8.30 back to 8.29. There is good spectroscopic (ESR) evidence for the formation of a species such as 8.30. In the absence of 8.29 bleaching (i.e., oxidation of polyphenolics) does take place, but it is considerably slower than the rate of bleaching in its presence.

The development and commercialization of 8.29-type catalysts in detergent formulations have an interesting history. The invention has been covered by more than 30 patents and estimated to have incurred a developmental cost of more than 150 million dollars. The performance of the detergent and its effect on fabrics have been the subject of litigation between Unilever and Proctor & Gamble. The current status is that the use of manganese catalyst-containing detergents is restricted to special areas.

8.7   ENGINEERING AND SAFETY CONSIDERATIONS

Liquid-phase oxidation reactions are generally carried out in a continuous stirred tank reactor by blowing pure oxygen or air at a pressure higher than the reactor operating pressure. The oxygen concentration in the gas phase at the top of the reactor must be below the flammability limit. As a further safety, the oxygen concentration is maintained below 5%. Because of these restrictions the volumetric production is generally low. Also, to reduce the gas and hydrocarbon vapor accumulation, these reactions are designed with very low headspace. When air is used as the oxidizing agent, the nitrogen carries liquid reactants, solvent, and products out of the reactor through the reactor vent, which needs to be recovered in a scrubber.

Organic peroxides are generally very unstable and can decompose spontaneously and explosively under thermal and mechanical stress. Such decomposition may be caused by shock, impact, friction, or by the catalytic effect of impurities. To reduce hazards involved during transportation and handling, they are desensitized by the addition of inert inorganic solids or liquids like water, halogenated hydrocarbons, etc. A solution of 70% t-butyl hydroperoxide has a self-accelerating decomposition temperature of 88°C. The acidity of hydroperoxides is greater than that of the corresponding alcohols. In many cases salts can be prepared that can be isolated and purified from the reaction mixture.

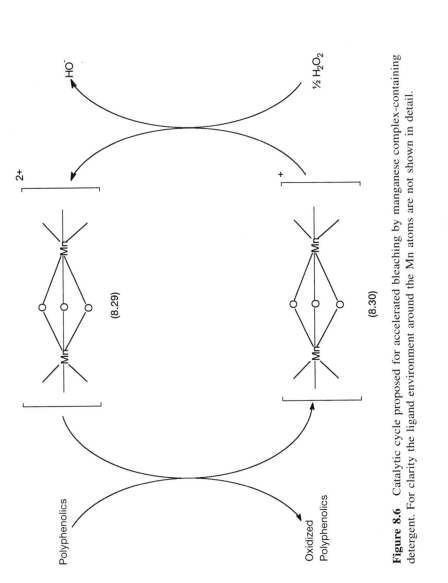

Figure 8.6 Catalytic cycle proposed for accelerated bleaching by manganese complex-containing detergent. For clarity the ligand environment around the Mn atoms are not shown in detail.

PROBLEMS

1. Give four examples of industrially important oxidation reactions that employ homogeneous catalysts. What are the fundamental mechanistic characteristics of these reactions?

Ans. Ethylene to acetaldehyde, cyclohexane to adipic acid, p-xylene to terephthalic acid, and propylene to propylene oxide. For the first, redox chemistry of palladium and nucleophilic attack on coordinated ligand. For the second and third, radical chain, and for the fourth, oxygen atom transfer.

2. Write the overall stoichiometries for ethylene to acetaldehyde, ethylene to vinyl acetate, and butene to methyl ethyl ketone by Wacker-type reactions. In each case write separately all the individual redox steps.

Ans. Sum reactions 8.1 to 8.3. Write reactions similar to 8.1 with HOAc in place of water, and butene in place of ethylene. All the other steps remain the same.

3. Identify reaction steps in Fig. 8.2 that illustrate the concepts of (a) the *trans* effect, (b) nucleophilic attack on a coordinated ligand, (c) insertion of alkene into an M–H bond, and (d) intramolecular electron transfer.

Ans. (a) 8.2 to 8.3. For a discussion of the *trans* effect, see *Inorganic Chemistry*, by D. Shriver, P. Atkins, and C. H. Langford, W. H. Freeman, New York, 1994. (b) 8.3 to 8.4. (c) 8.6 to 8.7. (d) 8.7 to 8.8.

4. What would be the isotopic composition of acetaldehyde in the following reactions, and what would be the significance of the results? (a) C_2D_4 + $PdCl_4^{2-}$ + H_2O; (b) C_2H_4 + $PdCl_4^{2-}$ + H_2O^{18}; (c) $C_2D_2H_2$ + $PdCl_4^{2-}$ + H_2O.

Ans. (a) CD_3CDO, vinyl alcohol tautomerization not the route to acetaldehyde; (b) CH_3CHO^{18}, oxygen atom from water and not air; (c) $CH_{3-x}D_xCH(D)O$ consistent with β-abstraction followed by insertion into the M–H(D) bond.

5. Based on Fig. 8.2, suggest tentative catalytic cycles for the oxidation of (a) ethylene to vinylacetate (VA) and (b) 2-butene to methyl ethyl ketone (MEK).

Ans. (a) A mechanism as in 8.2 to give an intermediate similar to 8.6 with coordinated VA, followed by elimination of HCl and VA. No 8.7-type intermediate. (b) β-hydrogen abstraction from the OH group containing C atom to give 8.6-type intermediate followed by 8.7-, 8.8-type intermediates.

6. Write the radical chain reactions for the formation of the hydroperoxide shown in Fig. 8.1, Route (b). What will be the effect metal ions such as Co^{2+} on this reaction?

Ans. Reactions 8.11–8.14; Co^{2+} will decompose the hydroperoxide to give the corresponding alcohol and dioxygen.

7. To what extent are the following statements true or false: (a) E° values for $M^{n+}/M^{(n+1)+}$ half-cells are good indicators for the overall product yields of auto-oxidation reactions. (b) In the radical-chain mechanism of auto-oxidation, only the organic hydroperoxide take part in electron transfer to and from the metal ion. (c) In vigorously purified hydrocarbon the metal ion must be present for generation of hydroperoxide. (d) Metal–dioxygen complexes enhance the rates of radical chain initiation and propagation steps. (e) In the oxidation of cyclohexane, under steady-state conditions (rates of initiation and termination equal), the molar ratio of cyclohexanone to cyclohexanol is 1:1.

Ans. (a) False. E° values are good indicators of catalytic efficiency only for the initiation steps. (b) False. The organic substrate (e.g., p-xylene, toluene, etc.) donate electrons to Co^{3+}. (c) and (d) False. In the absence of a radical chain there is no evidence to show that metal ions help in the production of hydroperoxide. Similarly, there is no evidence to show that metal–dioxygen complexes are involved in the initiation or propagation steps. (e) False. Cyclohexanol is also formed by 8.19. The ratio under steady state is 1:2.

8. What thermodynamic data may be used to rationalize the observations: (a) Cobalt catalysts can initiate autoxidation of p-xylene but not p-toluic acid. (b) Oxidation of isobutane leads to t-butyl hydroperoxide rather than isobutyl hydroperoxide.

Ans. (a) E° of Co^{3+}/Co^{2+} is 1.82. For $Ar^{+\bullet}/Ar$, E° values must be <1.82 and >1.82 when Ar = p-xylene and p-toluic acid, respectively. (b) C–H bond energy of Me_3C–H is lower than the others.

9. Write equations to explain: (a) Use of $Br^-/Br^{\bullet}$ as co-catalyst in cobalt-catalyzed autoxidation. (b) Hydroxylation of hydrocarbon by Cyt P450.

Ans. (a) $RH + Br^{\bullet} \rightarrow R^{\bullet} + HBr$, $Co^{3+} + Br^- \rightarrow Co^{2+} + Br^{\bullet}$. (b) $RH +$ "Fe=O" (Fe^{5+} or Fe^{4+} with porphyrin radical cation) $\rightarrow ROH + Fe^{3+}$.

10. In the molybdenum-catalyzed epoxidation of alkenes with t-butyl hydroperoxide what would be the effect of: (a) Addition of external t-butanol in the reaction mixture; (b) use of $[MoO_2(EG)_2]^{2+}$ (EG = ethylene glycol) as the precatalyst rather than $Mo(CO)_6$.

Ans. (a) Inhibition of rate; (b) shorter induction time as molybdenum is already in the 6+ oxidation state.

11. What are the names and/or type of rearrangements shown in Fig. 8.1, and synthesis of caprolactam?

Ans. Carbonium ion (Baeyer–Villiger type) and Beckmann rearrangements.

12. What are the main products of metal-catalyzed autoxidation of methyl cyclohex-2-ene? Why is cyclohexene more susceptible to autoxidation than cyclohexane?

Ans. Methyl cyclohex-2-en-4-one and methyl cyclohex-2-en-4-ol. Allylic stabilization of R$^\bullet$.

13. Model complexes of V^{5+} of the type 8.22 react with alkenes to give epoxides. What kind of kinetic evidence is a sure indication of alkene coordination?

Ans. Plot of 1/rate versus 1/[alkene] is a straight line.

14. Ti^{4+}-catalyzed epoxidation of allyl alcohol with t-butyl hydroperoxide is about one thousand times faster than that of n-hexene under identical conditions. Why?

Ans. Allyl alcohol coordinates to titanium through the OH functionality. The rate-determining step is unimolecular (both hydroperoxide and substrate are coordinated to metal) for allyl alcohol but bimolecular (n-hexene does not coordinate) for n-hexene. The favorable entropy of activation for allyl alcohol gives a faster rate.

15. With a Mn^{3+} Schiff base complex as the epoxidation catalyst, from the point of view of selectivity, would an organic hydroperoxide or sodium hypochlorite be a better oxygen atom donor?

Ans. NaOCl. With hydroperoxide some radical decomposition is unavoidable because of the availability of Mn^{2+}/Mn^{3+}/Mn^{4+}, etc.

16. What do you expect to happen when: (a) 8.29 is reacted with catechol and the reaction is monitored by ESR spectroscopy? (b) A catalytic amount of 8.29 is added to a mixture of a water-soluble alkene and H$_2$O$_2$?

Ans. (a) No ESR signal for 8.29 due to strong antiferromagnetic coupling between two Mn^{4+}. On addition of catechol, reduction of one of the Mn^{4+} and a 16-line spectrum (Mn nuclear spin 7/2) due to Mn^{3+}, Mn^{4+} unit (see R. Mage et al., *Nature* **369**, 637–38, 1994). (b) Formation of epoxide.

17. A dinuclear ruthenium complex with both ruthenium in 2+ oxidation states catalyzes the oxidation of adamantane to hydroxy adamantane and alkene to epoxide by dioxygen. Suggest a possible mechanism.

Ans. 2Ru^{2+} + O$_2$ → Ru-O-O-Ru → Ru=OO=Ru. The oxo species oxidizes adamantane and alkene and goes back to Ru^{2+} (see R. Neumann et al., *Nature* **388**, 353–54, 1997).

BIBLIOGRAPHY

For all the sections

Books

Most of the books listed under Sections 1.3 and 2.1–2.3.4 contain information on aspects of catalytic oxidation reactions and should be consulted. Especially useful are the books by Parshall and Ittel, and *Metal Catalyzed Oxidations of Organic Compounds: Mechanistic Principles and Synthetic Methodology Including Biochemical Processes*, R. A. Sheldon and J. Kochi, Academic Press, New York, 1981.

Also see Section 2.4 of Vol. 1 of *Applied Homogeneous Catalysis with Organometallic Compounds*, ed. by B. Cornils and W. A. Herrmann, VCH, Weinheim, New York, 1996.

For Wacker oxidation, see *The Organic Chemistry of Palladium*, P. M. Maitlis, Academic Press, New York, 1971; *Palladium Catalyzed Oxidation of Hydrocarbons*, P. M. Henry, D. Reidel, Dordrecht, 1980.

Articles

Sections 8.1 and 8.2

R. A. Sheldon and J. Kochi, in *Advances in Catalysis*, ed. by D. D. Eley, H. Pines, and P. B. Weisz, Academic Press, New York, Vol. 25, 1976, pp. 272–413.

P. M. Henry, in *Advances in Organometallic Chemistry*, ed. by F. G. A. Stone and R. West, Academic Press, Vol. 13, 1975, pp. 363–452.

J. E. Backvall et al., *J. Am. Chem. Soc.* **101**, 2411–16, 1979.

R. Jira, in *Applied Homogeneous Catalysis with Organometallic Compounds* (see under books), Vol. 1, pp. 374–93.

Sections 8.3 to 8.5

R. A. Sheldon and J. Kochi, in *Advances in Catalysis*, ed. by D. D. Eley, H. Pines, and P. B. Weisz, Academic Press, New York, Vol. 25, 1976, pp. 272–413.

D. M. Roundhill, in *Homogeneous Catalysis with Metal Phosphine Complexes*, ed. by L. H. Pignolet, Plenum Press, New York, 1983, pp. 377–403.

R. W. Fischer, in *Applied Homogeneous Catalysis with Organometallic Compounds* (see under books), Vol. 1, pp. 439–64.

Section 8.6

H. Mimoun et al., *J. Am. Chem. Soc.* **108**, 3711–18, 1986.

R. A. Sheldon, in *Applied Homogeneous Catalysis with Organometallic Compounds* (see under books), Vol. 1, pp. 411–23.

R. A. Sheldon, *Chemtech.* **21**(9), 566–76, 1991.

M. G. Finn and K. B. Sharpless, in *Asymmetric Synthesis*, ed. by J. D. Morrison, Academic Press, New York, Vol. 5, 1985, pp. 247–308.

Section 8.7

W. P. Griffith, *Chem. Soc. Review* **21**(3), 179–85, 1992.

References given in answers to Problems 16 and 17.

Section 8.8

N. I. Sax, *Dangerous Properties of Industrial Materials*, 5th edn, Van Nostrand Reinhold Co., New York, 1979.

CHAPTER 9

ASYMMETRIC CATALYSIS

9.1 INTRODUCTION

In Chapter 1, we have seen a few examples of the use of homogeneous catalysis for the manufacture of chiral molecules (Table 1.1). In this chapter we discuss such applications in more detail. Readers unfamiliar with stereochemical definitions and conventions should first consult the general references given at the end of this chapter. Depending on the number of asymmetric centers, chiral molecules have two or more optical isomers. Most biological reactions and processes are stereospecific, and in general, only one of the optical isomers of a chiral compound is responsible for the desired biological activity. In this connection, it is worth remembering that amino acids—the building blocks of proteins—occur only in the *levo* form in all biological systems.

A racemic mixture of a pharmaceutical or an agrochemical consists of equal amounts of two enantiomers, only one of which is biologically active. The other one may have no biological effect on the desired biological reaction. In fact, in certain instances it may have highly adverse side effects. The biological effects of chiral drugs may be extremely complicated. The drug thalidomide is a good example. Originally it was thought that the *S* isomer 9.1 was teratogenic (i.e., the isomer responsible for deformities in newborn babies). However, it turns out that single isomers of thalidomide undergo rapid interconversion in plasma and in human volunteers.

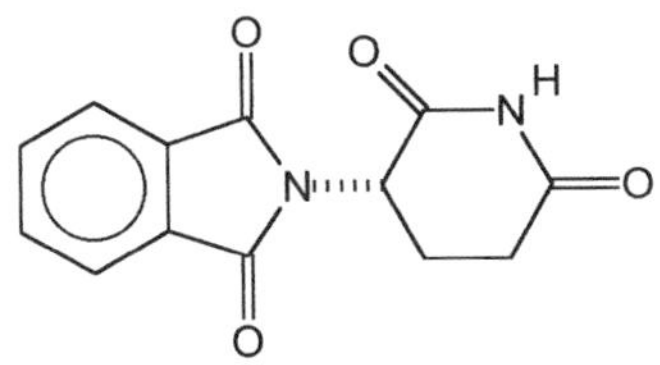

(S-THALIDOMIDE)

(9.1)

This observation appears to make the question of racemic thalidomide versus a single isomer drug moot. Pharmacological studies based on racemates rather than on pure enantiomers can often be misleading. These considerations have given rise to stringent guidelines for the registration of chiral drugs, where two enantiomers of a chiral drug will have to be separately tested for their biological activities.

In principle, three approaches may be adopted for obtaining an enantiomerically pure compound. These are resolution of a racemic mixture, stereoselective synthesis starting from a chiral building block, and conversion of a prochiral substrate into a chiral product by asymmetric catalysis. The last approach, since it is catalytic, means an amplification of chirality; that is, one molecule of a chiral catalyst produces several hundred or a thousand molecules of the chiral product from a starting material that is optically inactive! In the past two decades this strategy has proved to be extremely useful for the commercial manufacture of a number of intermediates for biologically active compounds. A few recent examples are given in Table 9.1.

With the exception of S-metolachlor, all the molecules listed under the column Final Target are used in pharmaceutical formulations. Dilitiazem is a Ca^{2+} antagonist, while Cilazapril is an angiotensin-converting enzyme inhibitor. Levofloxacin is an antibacterial, and cilastatin is used as an in vivo stabilizer of the antibiotic imipenem. S-metolachlor is a herbicide sold under the trade name of DUAL MAGNUM. Although the structures of the final targets are more complex than those of the intermediates, enantioselective syntheses of the intermediates are the most crucial steps in the complex synthetic schemes of these molecules.

The other approach, resolution of a racemic mixture, is also used in many industrial processes. Here physical resolution (i.e., resolution by crystallization of a diastereomeric derivative) is in general the favored method. However, kinetic resolution in principle can also be used for separating out two enantiomers. Although kinetic resolution involving a soluble metal complex catalyst has yet to be widely practiced, the potential importance of such an approach is significant (see Section 9.3.5).

9.2 GENERAL FEATURES OF CHIRAL LIGANDS AND COMPLEXES

The design of a chiral catalyst often starts with an achiral metal complex that exhibits some activity in the desired catalytic reaction. Once such a complex is identified, the ligands are modified in such a way that a chiral environment is created. Further fine tuning of the ligand and other reaction conditions may then be undertaken to optimize the enantioselectivity of the reaction.

The ligands that we are going to discuss in this chapter are in most cases no different from the general categories that we have already encountered in the previous chapters. They can be broadly divided into two types. The first type has "hard" donor atoms such as nitrogen and/or oxygen, and could be monodentate or chelating depending upon the nature of the reaction. Strictly speaking nitrogen as a donor atom is at the borderline between "hard" and "soft." Here for simplification it is classified as a "hard" donor atom. These ligands can complex and stabilize metal ions in high oxidation states. The second type is almost exclusively based on chelating phosphines. As should be evident from discussions in the earlier chapters, the first type of ligand with metal ions such as Ti^{4+}, Os^{8+}, Rh^{3+}, Mn^{3+}, etc. are used for reactions such as epoxidation, alkene dihydroxylation, and cyclopropanation. The second category of ligands with Rh^+, Ru^{2+}, Ir^+, Ni^0 find uses in hydrogenation, isomerization, hydrocyanation, hydroformylation, etc. Some of the ligands and metal complexes that have been industrially used or actively evaluated are shown in Figs. 9.1 and 9.2. The ligands and complexes shown in Fig. 9.1 belong to the first category, while those in Fig. 9.2 belong to the second one.

Complex 9.7, where a simple chiral Schiff base is used, is probably the first example of an asymmetric homogeneous catalyst. It catalyses the asymmetric cyclopropanation of alkenes by carbene addition with low enantioselectivity. The potential of this and other similar chiral Schiff base complexes of copper were originally explored for the manufacture of synthetic pyrethroids, a class of highly active pesticides. Cyclopropane rings with asymmetric centers are integral parts of the molecular structures of synthetic pyrethroids. In recent years tremendous improvement in enantioselectivity (e.e. > 99%) has been achieved for cyclopropanation reactions with copper complexes of the chiral bisoxazoline ligand 9.8. In these complexes bidentate chelation to copper takes place through the nitrogen atoms. Intermediate 9.6 of Table 9.1 is also synthesized by asymmetric cyclopropanation.

The macrocyclic chemistry of tetradentate Schiff base complexes has been known for long time. However, the successful use of such a complex as an enantioselective catalyst in epoxidation reactions is a relatively recent finding. In these reactions complex 9.9 or an analogue is used. One of the possible routes for the synthesis of intermediate 9.2 of Table 9.1 involves the use of a similar catalyst. While complex 9.9 works well with unfunctionalized alkenes, for the epoxidation of allylic alcohols, dialkyl tartarates, 9.10, are the preferred ligands. As we shall see, the mechanisms of epoxidation in these two cases are different. Also for the tartarate-based system titanium is the metal of choice (see Section 9.3.3).

The ligands in 9.7–9.9 are all based on N,O or N,N donor-atom-based chelating systems. In contrast, the ligands in 9.10–9.12 are all bidentate alkoxides, with O,O donor atoms. The chirality in binapthol, 9.11, arises from the fact that due to steric interactions the molecule does not have a plane of symmetry. The two naphthyl rings lie in different planes. The complex 9.12 has binaphthol as the ligand. Complexes such as 9.12 are used as catalysts in nitroaldol condensation reaction (see Section 9.5.4).

The chiral motifs of both 9.10 and 9.11 have been exploited to make chiral phosphines. Thus 9.13, abbreviated as DIOP, and the phosphine ligand in 9.16,

TABLE 9.1 Intermediates for Biologically Active Compounds

Asymmetric reaction and status	Intermediate	Final target
Epoxidation or dihydroxylation; alternative routes are under evaluation	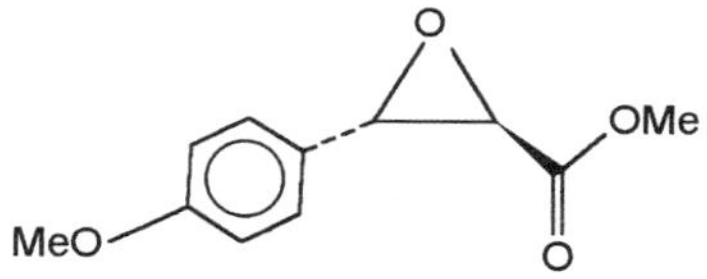 (9.2)	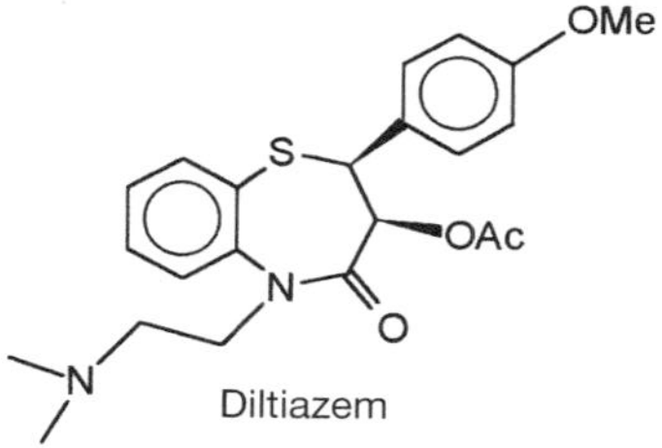 Diltiazem
Hydrogenation of C=C; commercialized by Hoffmann–La Roche, Ltd.	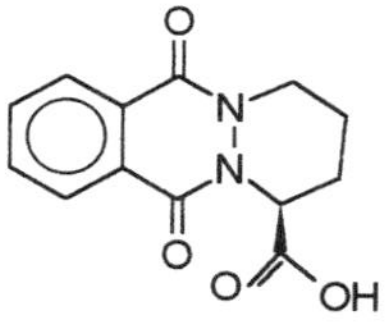 (9.3)	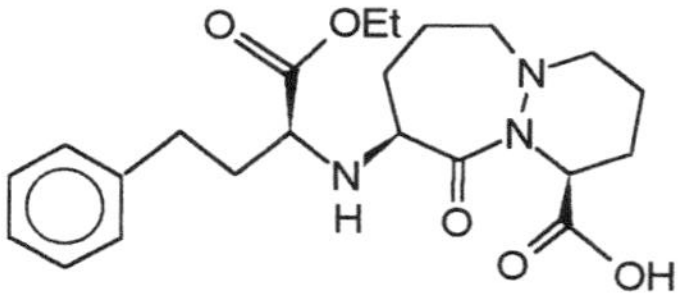 Cilazapril
Hydrogenation of C=N; commercialized by Novartis Services, Ltd.	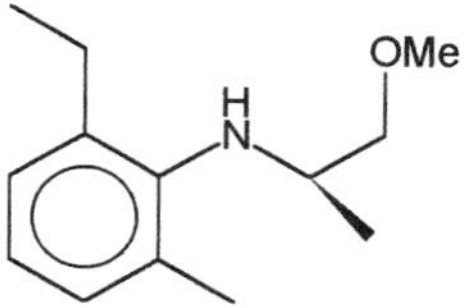 (9.4)	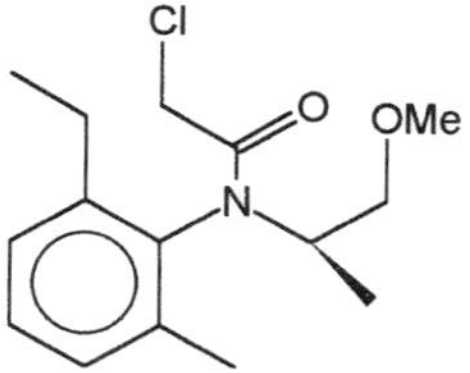 (S)-Metolachlor

Hydrogenation of C=O; commercialized by
Daiichi Pharmaceuticals/Ikasago Industrial
Corporation

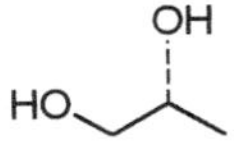

(9.5)

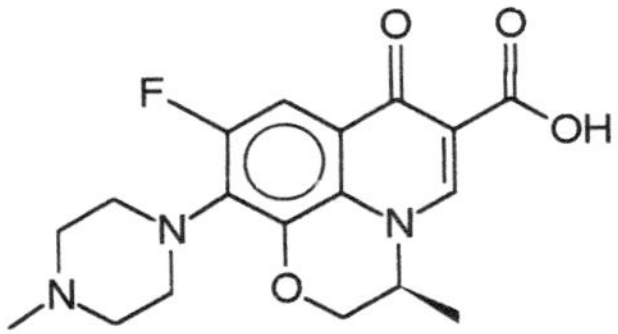

Levofloxacin

Asymmetric cyclopropanation; commercialized
by Sumitomo Chemical Co., Ltd., and Merck
& Co.

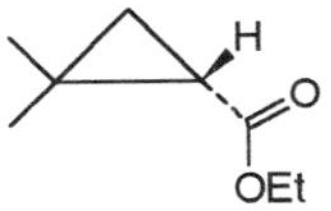

(9.6)

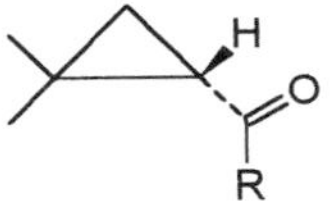

Cilastatin

$R =$

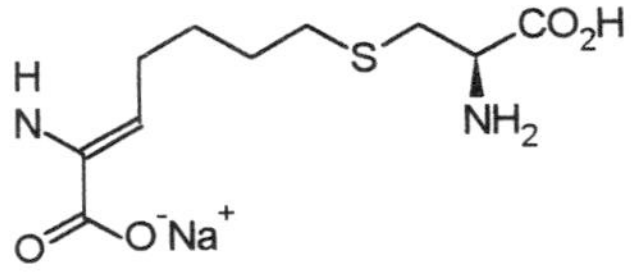

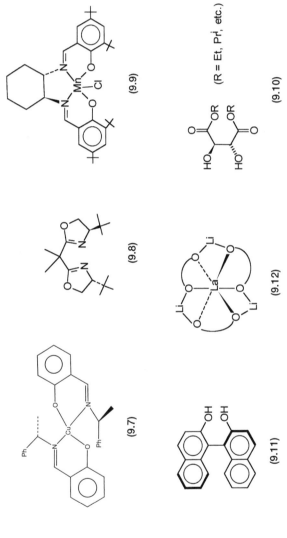

Figure 9.1 Chiral ligands and metal complexes with "hard" donor atoms. The complexes of these ligands with metal ions in relatively high oxidation states are used in asymmetric epoxidation, cyclopropanation, and nitroaldol condensation reactions.

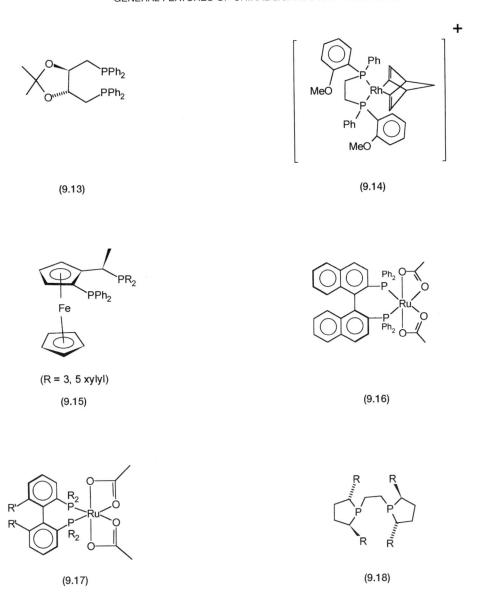

(9.13)

(9.14)

(R = 3, 5 xylyl)

(9.15)

(9.16)

(9.17)

(9.18)

Figure 9.2 Chiral chelating phosphines and their metal complexes. These ligands and complexes are used in asymmetric hydrogenation of C=C, C=O, and C=N double bonds.

commonly referred to as BINAP, are synthesized from tartaric acid and bi-naphthol, respectively. In the metal complexes of BINAP, the dihedral angle between the naphthalene rings could be as large as 75°. Intermediate 9.5 is manufactured by the hydrogenation of the corresponding prochiral ketone using 9.16 as the catalyst.

Complex 9.14 was one of the first asymmetric hydrogenation catalysts. Monsanto developed it for the manufacture of L-DOPA (see Table 1.1). The chelating *bis*-phosphine in this case is called DIPAMP. Note that the chirality of DIPAMP is located on the phosphorus atoms. The rhodium complex of DIOP, like that of DIPAMP, gives high enantioselectivity in the hydrogenation of prochiral alkenes with α-acetamido functionality. The iridium complex of ligand 9.15 is used in the manufacture of 9.4. This is brought about by the asymmetric hydrogenation of the corresponding prochiral imine. To date this is the only reported example of a large-scale industrial process ($>10^4$ ton/year) that involves the use of an iridium complex as a homogeneous catalyst.

Complex 9.17 has obvious similarities with 9.16 in the sense that the chirality of the ligand arises from the nonplanarity of the molecule. The nonplanarity is a result of the presence of bulky R groups. Intermediate 9.3 is manufactured industrially by hydrogenating the corresponding prochiral cyclic alkene using 9.17 (R=*p*-tolyl) as the catalyst. The chelating phosphine 9.18 has also been found to induce high activity and enantioselectivity in the hydrogenation of a variety of prochiral substrates.

9.3   MECHANISMS AND CATALYTIC CYCLES

Although a large number of asymmetric catalytic reactions with impressive catalytic activities and enantioselectivities have been reported, the mechanistic details at a molecular level have been firmly established for only a few. Asymmetric isomerization, hydrogenation, epoxidation, and alkene dihydroxylation are some of the reactions where the proposed catalytic cycles could be backed with kinetic, spectroscopic, and other evidence. In all these systems kinetic factors are responsible for the observed enantioselectivities. In other words, the rate of formation of one of the enantiomers of the organic product is much faster than that of its mirror image.

Note that in a number of the molecular structures shown in Figs. 9.1 and 9.2, there is a common element of symmetry. Since all of them are chiral, they do not possess any alternating axis of symmetry, but may have a simple axis of symmetry. As can be easily seen, even from the two-dimensional drawings, 9.8, the ligand of 9.9, 9.10, 9.11, 9.13, and 9.18 have C_2 axes of symmetry.

The interaction of a prochiral molecule with a chiral homogeneous catalyst results in the formation of *diastereomeric* intermediates and transition states. High enantioselectivity is obtained if for the rate-determining step out of the many possible diastereomeric transition states, one is energetically favored. In such a situation the reaction follows mainly this path. Other possible pathways that cause dilution of optical purity are avoided. A fundamental point to note is that diastereomers, unlike enantiomers, need not have identical energies. The presence of a C_2 axis of symmetry in these ligands makes some of the possible diastereomeric transition states structurally and energetically equivalent. As the total number of energetically different transition states reduces, the chances of one of them being energetically favored increases.

A simple calculation would show that a lowering of about 2.8 kcal in the free energy of activation under ambient conditions increases the rate constant by about a hundred times. Other conditions being equal, if this happens to one specific diastereometeric transition states out of many, one enantiomer of the product would be selectively formed. Ideally we would like to predict the kind of ligand environment that may induce an energy difference of this order between the various possible diastereomeric transition states. Unfortunately, to date there is no reliable a priori method that could accurately calculate energies of all possible transition states, and tell us what precise ligand environment would be required to bring this about. Development of an efficient asymmetric catalyst consequently remains a matter of creating a large library of experimental data backed by theoretical considerations.

9.3.1   Mechanism of Asymmetric Hydrogenation

Reaction 9.1 has been extensively studied to establish the mechanism of asymmetric hydrogenation. The catalytic cycle proposed for the asymmetric hydrogenation of the methyl ester of α-acetamido cinamic acid with 9.14 as the precatalyst is shown in Fig. 9.3. As mentioned earlier, this reaction is one of the early examples of industrial applications of asymmetric catalysis for the manufacture of L-DOPA (see Table 1.1).

(9.1)

(S) isomer

Catalyst = (9.14) with R, R DIPAMP

The following points are to be noted about the catalytic cycles. First, the precatalyst 9.14 reacts with dihydrogen to give norborane and generate two diastereomeric intermediates, 9.19 and 9.20. In both these intermediates the O atom of the carbonyl group of the acetamido functionality occupies a ligand site and coordinates to the rhodium center. In other words, the substrate acts as a chelating ligand by coordinating through the olefinic double bond and the oxygen atom. The C=C coordination to the metal center can take place through two different faces (the so-called Re and Si faces) to give diastereomers 9.19 and 9.20. In these two complexes the alkene group containing parts have a mirror-image relationship with each other, but the chirality of the chelating

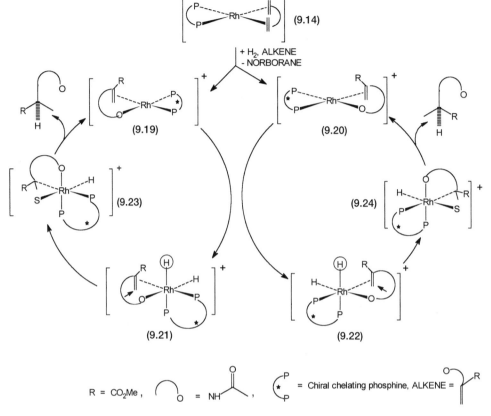

Figure 9.3 Catalytic cycles for the asymmetric hydrogenation of α-acetamido methyl acrylate (or cinamate). For clarity the detailed structure of the organic substrate is not shown. In 9.21 and 9.22 for ease of identification the carbon atom to which the metal hydride is transferred is marked by an arrow and the hydride is circled. Note that, excepting the chelating chiral phosphine, the stereochemistries around the rhodium in the left- and right-hand cycles have mirror-image relationships.

phosphine is obviously the same. For this reason 9.19 and 9.20 have a *dia-stereomeric* rather than an *enantiomeric* relationship.

Oxidative addition of dihydrogen to 9.19 and 9.20 produces the intermediates 9.21 and 9.22, respectively. Insertion of the alkene into the Rh–H bonds produces diastereomers 9.23 and 9.24. It is important to note that the coordination site vacated by the hydride ligand, circled for easy identification, is taken up by the O atom of the carbonyl functionality of the acetamido group, and a solvent molecule occupies the original position of the O atom.

This has the important effect of fixing the opposite of the initial chirality of the prochiral carbon. In other words Re and Si faces bonded to Rh generate *S* or *R* chirality, respectively. This is shown schematically in Fig. 9.4. The detailed

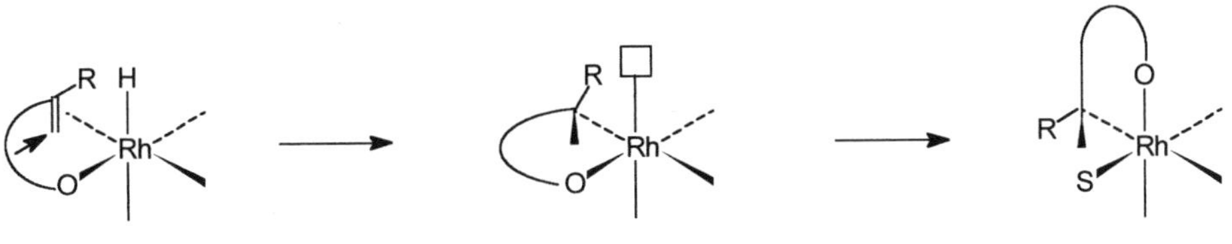

(9.25)   (9.26)   (9.27)

Figure 9.4 Proposed reaction sequence to explain why the stereochemistry of the chiral carbon atom is opposite to what is expected from a simple hydride transfer. Note that the stereochemistry of the carbon atom in 9.26 is changed to that of 9.27 by the breakage of the Rh–O bond so that rotation round the Rh–C bond becomes possible. The oxygen atom then occupies the empty coordination site marked by the square, and the Rh–O bond is formed again.

mechanism for the conversion of 9.25 to 9.27 that results in chirality inversion is not shown in the catalytic cycles of Fig. 9.3. For clarity, in Fig. 9.3 conversions of 9.21 and 9.22 to 9.23 and 9.24, respectively, have been shown in single steps. Finally, 9.23 and 9.24 undergo reductive elimination of the hydrogenated products to complete the catalytic cycles and regenerate 9.19 and 9.20.

The overall enantioselectivity of the catalytic process obviously depends on the relative speed with which the left and right catalytic cycles of Fig. 9.3 operate. Oxidative addition of dihydrogen is found to be the rate-determining step. Therefore, the relative rates of conversion of 9.19 to 9.21 on the one hand and 9.20 to 9.22 on the other determine which enantiomer of the organic product would be formed preferentially. The reaction between 9.28 and α-acetamido methyl cinnamate has been monitored by multinuclear NMR, and both 9.19 and 9.20 have been identified. Depending on the stereochemistry of the chiral phosphine, one of these two diastereomers is preferentially formed.

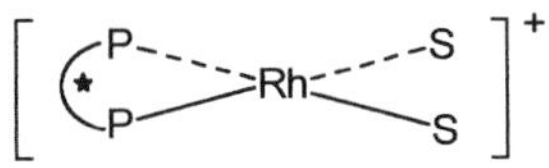

(9.28)

NMR data indicate the ratio of the concentrations of the major and the minor isomer to be approximately 10:1. Since the concentration of one of the isomers is almost ten times the other, if the rate constants for oxidative additions of dihydrogen are approximately the same, the major diastereomer should undergo conversion to the dihydride ten times faster. This, however, is *not* the mechanism for enantioselection. The mechanism of enantioselection is the much larger rate constant (~600 times) for the reaction between the *minor* isomer and dihydrogen!

The bulk of the evidence for the proposed mechanism comes from elegant kinetic studies on a variety of chelating phosphine-based catalytic systems. The reactions of the two diastereomers 9.19 and 9.20 with dihydrogen have also been separately studied by low-temperature NMR. The results of these studies are consistent with the above-mentioned mechanism for enantioselection. The X-ray structure of an analogue of the major isomer with a different chelating phosphine (S,S-chiraphos) has been reported. Using an achiral chelating phosphine, an analogue of 9.23 and 9.24 has also been identified by multinuclear NMR.

The detailed kinetic and thermodynamic data relating to the stabilities of 9.19, 9.20, and the activation energy barriers for the oxidative addition of dihydrogen can be summed up in the free-energy diagram shown in Fig. 9.5. The most interesting point of this diagram is the fact that the two free-energy profiles cross each other near the transition states. This is because the thermodynamically more stable major isomer has a transition state that is higher in energy than that of the minor isomer. Note that the free energies of the

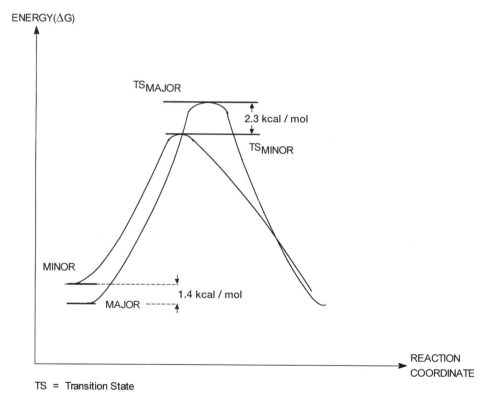

TS = Transition State

Figure 9.5 Free energy diagram for the oxidative addition of dihydrogen to the two diastereomers of [Rh(R,R DIPAMP)(S)]⁺, where S = methyl-(Z)-α-acetamidocinamate. The "major" and "minor" refer to intermediates 9.19 and 9.20, one of which has higher equilibrium concentration than the other. The one with higher equilibrium concentration is called the major, and the other is called the minor isomer.

products 9.21 and 9.22 are arbitrarily fixed. Their diasteromeric relationship means that they need not have identical energies. The important point is that the free energies of activation, rather than free energies of the starting complexes, determine the enantioselection process.

9.3.2 Asymmetric Isomerization and Mechanism

Like the synthesis of L-DOPA by asymmetric hydrogenation, the manufacture of L-menthol by Takasago Company is also one of the early examples of an industrial process where asymmetric isomerization is a key step. The desired isomerization reaction is one of the steps of the overall synthetic scheme. The synthesis of L-menthol from diethyl geranylamine is shown by 9.2. The formal electron pair pushing mechanism for the isomerization of the allylic amine to the enamine proceeds according to reaction 9.3.

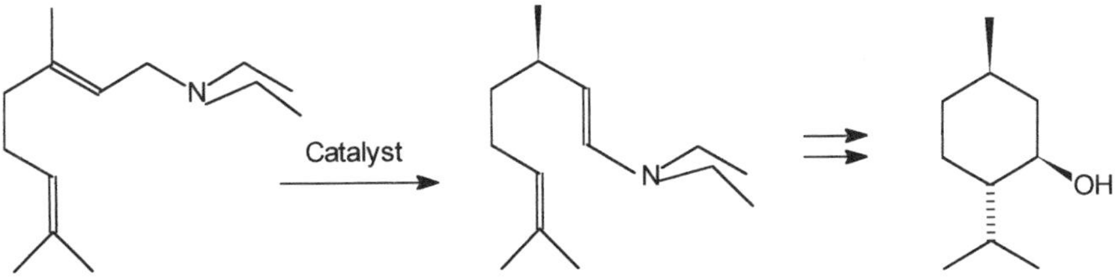

Catalyst = Analogue of (9.14) with S-BINAP

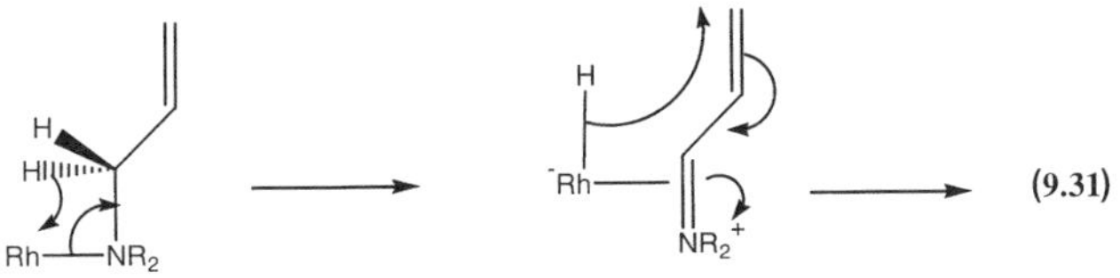

The enamine in 9.31 can coordinate to the metal center in three different ways. It can act as a monodentate ligand by coordinating either through the N atom lone pair or the double bond. It can also act as a chelating ligand. As we will see, the proposed catalytic intermediates in the catalytic cycle have all three different types of coordination.

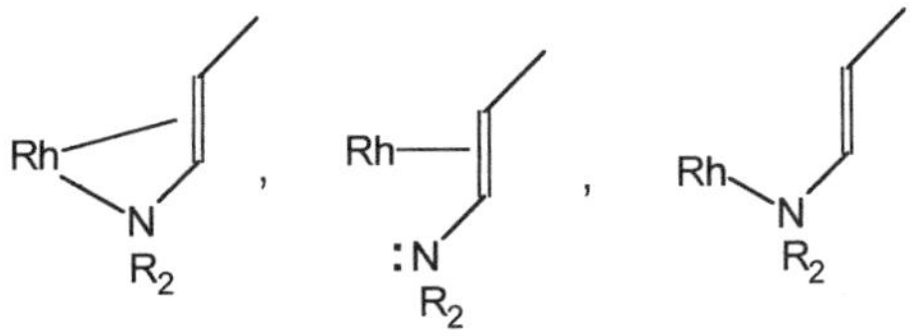

A simplified proposed catalytic cycle is shown in Fig. 9.6. The precatalyst, an analogue of 9.14 with S-BINAP, undergoes reaction 9.3 to generate 9.32, where the enamine acts as a chelating ligand. Note that in 9.32 asymmetric isomerization has already taken place. How this may come about will be dis-

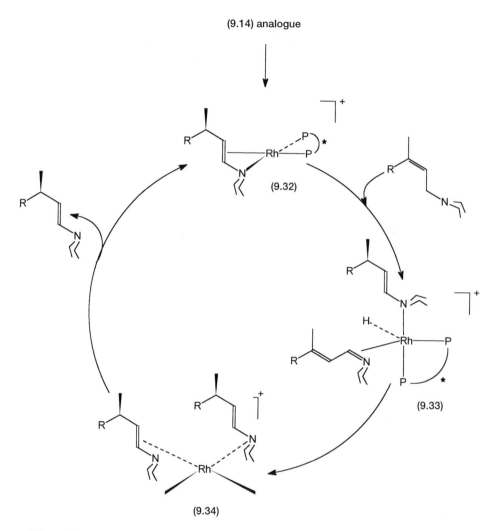

Figure 9.6 Catalytic cycle for the asymmetric isomerization of diethylgeranylamine. The precatalyst is an analogue of 9.14 with *S*-BINAP in place of DIPAMP. Conversions of 9.14 to 9.32 and 9.32 to 9.33 involve more than one step (see Problem 9). These are not shown for clarity.

cussed shortly. Although shown as a single step, the conversion of 9.32 to 9.33 involves two reactions. First, coordination by a molecule of the allyl amine 9.29 causes the enamine to act as a monodentate ligand that coordinates only through the nitrogen lone pair. This is followed by hydride abstraction from the α-carbon of 9.29 and the generation of the hydride complex 9.33.

Note that in 9.33 hydride abstraction generates quaternary nitrogen containing conjugated allyl amine ligand. This prochiral molecule coordinates to the metal through the carbon–nitrogen double bond. Conversion of 9.33 to 9.34 is the step where the H atom is added to the γ-carbon in a stereospecific manner. Presumably, the chiral environment around the metal center ensures that the hydride transfer on one enantioface is kinetically preferred. In 9.34 both the enamine molecules act as monodentate ligands. One of them coordinates through the alkene functionality, but the other uses the nitrogen lone pair. Chelation by any one of them leads to product formation, that is, elimination of the other enamine molecule as the product. This regenerates 9.32 and completes the catalytic cycle.

9.3.3   Asymmetric Epoxidation of Allylic Alcohols and Mechanisms

In epoxidation reactions allyl alcohol can act as a prochiral alkene. Enantiomerically pure glycidol isomers (see Table 1.1) may be used to make S-propanolol 9.61, a drug for heart disease and hypertension. The mechanistic details of the epoxidation reaction with V^{5+} and Mo^{6+} complexes as catalysts were discussed in Section 8.6. The basic mechanism of epoxidation reaction, the transfer of an oxygen atom from t-butyl hydroperoxide to the alkene functionality, remains the same.

The chiral precatalyst is a titanium species. It is generated by the in situ treatment of titanium isopropoxide with diethyl or diisopropyl tartarate. The relative amounts of $Ti(OPr^i)_4$ and the tartarate ester have a major influence on the rate of epoxidation and enentioselectivity. This is because the reaction between $Ti(OPr^i)_4$ and the tartarate ester leads to the formation of many complexes with different Ti:tartarate ratios. All these complexes have different catalytic activities and enantioselectivities. At the optimum Ti:tartarate ratio (1:1.2) complex 9.35 is the predominant species in solution. This gives the catalytic system of highest activity and enantioselectivity. The general phenomenon of rate enhancement due to coordination by a specific ligand, with a specific metal-to-ligand stoichiometry, is called *ligand-accelerated catalysis*.

(9.35)

Before discussing the structural evidence for the precatalyst 9.35, we quickly go through the proposed mechanism of epoxidation. The precatalyst 9.35 reacts with one mole each of allyl alcohol and t-butyl hydroperoxide to give 9.36, where two alkoxide ligands on the same Ti atom are substituted according to reaction 9.4.

$$(9.35) \; + \quad \text{(allyl alcohol, OH)} \quad + \quad \text{(}t\text{-Bu–OOH)} \quad \rightleftharpoons \quad (9.36) \; + \; 2\,R'OH \qquad (9.4)$$

The proposed catalytic cycle and the structure of 9.36 are shown in Fig. 9.7. Nucleophilic S_N2-type attack by the olefin to the distal oxygen atom produces the epoxy alkoxide. The chiral environment around the Ti atom ensures that the allyl alcohol is oriented in such a way that O atom transfer takes place only on one particular enantioface. The discrimination between the two possible faces is "stereoelectronic" rather than "steric" in nature. The epoxy alkoxide is then replaced by allyl alcohol to give the epoxy alcohol and 9.37. The latter can react with more t-butyl hydroperoxide to regenerate 9.36.

The evidence for the proposed mechanism comes from kinetic, spectroscopic (multinuclear NMR), X-ray structure, and theoretical calculations. The kinetic rate law under optimum catalytic conditions is very complex. Under pseudo-first-order conditions, where the concentrations of both 9.35 and the hydroperoxide are much greater than that of allyl alcohol, the rate expression 9.5 is obeyed. In this expression the inhibitor alcohol is an inert alcohol such as isopropanol or t-butanol that is deliberately added to slow down the reaction for convenient rate measurements. The inert alcohol acts as an inhibitor, since it competes with both hydroperoxide and allyl alcohol for coordination to the Ti center. Note that expression 9.5 is consistent with the formation of an intermediate like 9.36, before the rate-determining oxygen atom transfer step.

$$\text{Rate} = k' \, [\text{allyl alcohol}] \, [\text{Hydroperoxide}] \, [(9.35)] \, / \, [\text{Inhibitor alcohol}]^2 \qquad (9.5)$$

$$2\,Ti(OR)_4 \; + \; 2\,\text{Tartarate} \;\rightleftharpoons\; [Ti(OR)_2\,(\text{Tartarate})]_2 \; + \; 4\,ROH \qquad (9.6)$$

The equilibrium as shown by 9.6 has been observed by NMR, and the product formed in solution has been shown to be dimeric in nature by molecular weight measurements and multinuclear NMR. The proposed structure, 9.35, has two inequivalent CO_2R groups. One of these is bonded to the Ti center, while the other is free. X-ray structures of several Ti–tartarate complexes have also been determined, and they all exhibit the general structural features of 9.35. In other words, there is good evidence to suggest that the solid-state structure of 9.35 is retained in solution, and this species indeed is the true precatalyst.

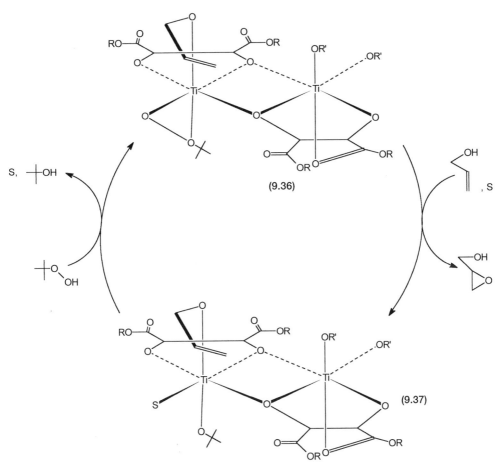

Figure 9.7 Catalytic cycle for asymmetric epoxidation of allyl alcohol with 9.35 as the precatalyst. The precatalyst is generated in situ and undergoes conversion to 9.36 in the presence of allyl alcohol and *t*-butyl hydroperoxide. S is a solvent molecule. Conversion of 9.36 to 9.37 involves more than one step. This is not shown for clarity (see Problem 10).

9.3.4 Asymmetric Epoxidation of Alkenes Other Than Allyl Alcohols

It is clear that in the asymmetric epoxidation of allyl alcohols, coordination by the OH functionality to the titanium center plays a crucial role. Such a coordination has a favorable effect on the entropy of activation, that is, the rate constant (see Question 14 of Chapter 8). It also helps to orient only one of the two possible enantiofaces for a facile oxygen atom transfer. With alkenes that do not have any such functional groups, the titanium tartarate system gives poor enantioselectivities. The precatalyst that has been found to exhibit remarkably high enantioselectivities for such alkenes is a planar manganese complex 9.38A with an optically active Schiff-base ligand. In these reactions, instead of Bu^{t}OOH, iodosyl benzene (PhIO) or NaOCl is used as the oxygen atom donor.

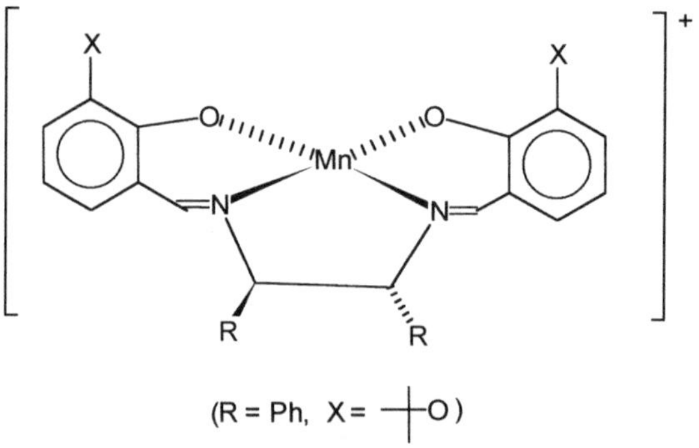

(R = Ph, X = —$\dagger$O)

(9.38A)

The proposed mechanism for epoxidation with 9.38B as the catalyst is shown in Fig. 9.8. The oxidation state of manganese in 9.38A and 9.38B is three. The oxygen donor NaOCl or PhIO oxidizes Mn^{3+} to Mn^{5+}, and an oxo complex such as 9.39 is produced. Reaction of 9.39 with the alkene produces the chiral epoxide and regenerates 9.38B. While there is enough evidence for the basic mechanism and the involvement of a catalytic intermediate such as 9.39, there is some controversy about the details of the oxygen atom transfer from 9.39 to the alkene.

Two alternative mechanisms have been proposed. A direct substrate attack at the oxo ligand with concerted or sequential C–O bond formation is possible. Alternatively, the substrate may attack at both the metal and oxo centers to generate an oxametallocyclic intermediate. These two alternatives are shown in Fig. 9.9. Finally, note that in CytP450-catalyzed hydroxylation and epoxidation, an iron porphyrin intermediate of the type of 9.39 is involved.

9.3.5 Asymmetric Hydrolysis of Epoxides

A recent discovery that has significantly extended the scope of asymmetric catalytic reactions for practical applications is the metal-complex-catalyzed hydrolysis of a racemic mixture of epoxides. The basic principle behind this is *kinetic resolution*. In practice this means that under a given set of conditions the two enantiomers of the racemic mixture undergo hydrolysis at different rates. The different rates of reactions are presumably caused by the diastereomeric interaction between the chiral metal catalyst and the two enantiomers of the epoxide. Diastereomeric intermediates and/or transition states that differ in the energies of activation are presumably generated. The result is the formation of the product, a diol, with high enantioselectivity. One of the enantiomers of

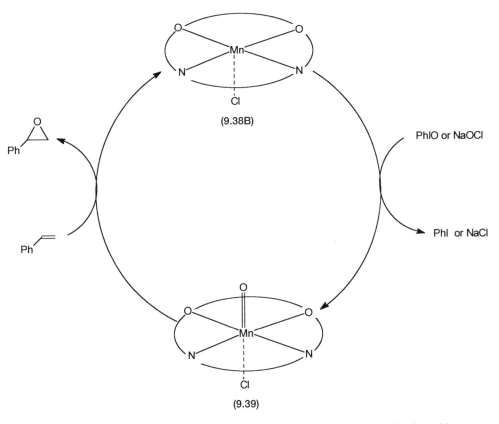

Figure 9.8 Catalytic cycle for the epoxidation of unfunctionalized alkenes with a chiral Schiff base complex of manganese as the catalyst.

the starting epoxide is left more or less untouched, because its rate of hydrolysis is much less compared to that of the other enantiomer. The overall reaction is shown by 9.7. The maximum theoretical yield of the diol is obviously half that of the starting epoxide.

$$(\pm) \quad \text{[epoxide]} \quad + \; H_2O \quad \xrightarrow[\text{catalyst}]{\text{chiral}} \quad \text{[epoxide]} \quad + \quad \text{[diol]} \tag{9.7}$$

Asymmetric hydrolysis has several specific advantages to offer. First of all it uses water as one of the reagents. Water is cheap, safe, and environmentally benign! Second, chiral 1,2 diols are versatile building blocks for complex organic molecules. Finally, asymmetric catalytic epoxidation does not work for alkenes such as propylene. However, by this method a racemic mixture of

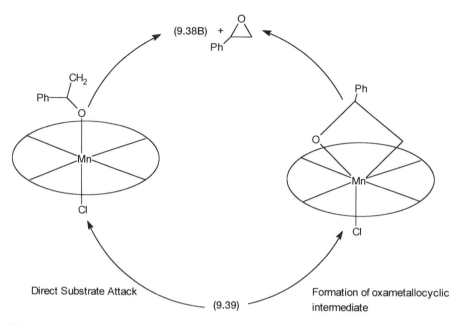

Figure 9.9   Two alternative pathways for the oxygen atom transfer step. The right-hand path involves formation of an oxametallocycle, whereas the left-hand path does not involve any such intermediate.

propylene oxide could be hydrolyzed with full theoretical yield and 98% enantioselectivity.

The precatalyst used in these water-based kinetic resolution reactions is the cobalt Schiff-base complex 9.40. Its structural similarity to the asymmetric epoxidation catalysts 9.38A and 9.38B is to be noted. In the actual catalytic system 9.40 is activated with small amounts of acetic acid and air to give a cobalt(III) complex where CH_3CO_2 and H_2O are additional ligands. The mechanistic details of this reaction are as yet unknown.

(9.40)

9.4 ASYMMETRIC DIHYDROXYLATION REACTION

Osmium-mediated dihydroxylation of carbon–carbon double bonds with OsO_4 is a classic reaction that can be made catalytic by using co-oxidants such as Bu^tO_2H or N-methylmorpholine N-oxide. For asymmetric dihydroxylation (ADH) reactions the co-oxidant of choice is the water-soluble $[K_3Fe(CN)_6]$ salt. As shown in Fig. 9.10, in a biphasic system consisting of water and a water-immiscible organic solvent, OsO_4 reacts with the alkene to produce the glycolate 9.41. Coordination by a chiral ligand L to osmium ensures that this step is enantioselective. Owing to the solubility properties of OsO_4, 9.41, alkene, and L, this step occurs in the organic phase.

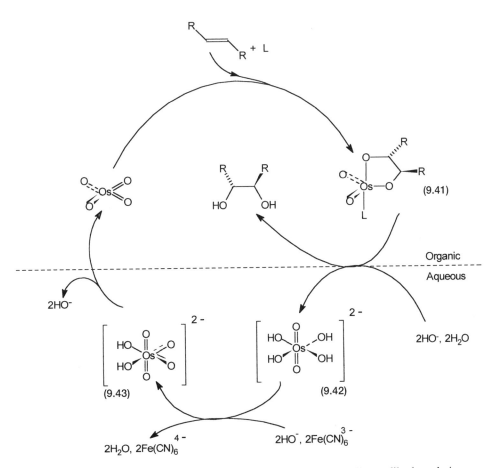

Figure 9.10 Catalytic cycle for OsO_4-catalyzed asymmetric alkene dihydroxylation. The dashed line represents the phase boundary between the organic and the aqueous phase. L is the chiral ligand, e.g., 9.44.

Base-catalyzed hydrolysis of 9.41 at the water–organic interface produces 9.42 and the optically active diol. Oxidation of 9.42 to 9.43 by ferricyanide takes place in the aqueous layer. The complexes 9.42 and 9.43 have only oxo and hydroxo groups and no coordinated organic ligands. Both of these are doubly charged anions. Consequently they have little or no solubility in nonpolar organic solvents. The oxidation states of osmium in these two complexes are, however, different. In 9.42 osmium is in the 6+ oxidation state, while in 9.43 it is in the 8+ oxidation state. Regeneration of OsO_4 from 9.43 by the loss of two HO^- ions completes the catalytic cycle.

With other organic-solvent-soluble co-oxidants such as NMO, a trioxo glycolate complex, with osmium in the 8+ oxidation state, is formed by the oxidation of 9.41. In other words, in the organic solvent apart from OsO_4, another Os^{8+} species is present. This complex, since it is soluble in the organic solvent, can initiate a secondary catalytic cycle with poor enantioselectivity. The advantage of using $[Fe(CN)_6]^{3-}$ as the co-oxidant is that in the organic solvent the only oxidant with osmium in the 8+ oxidation state is OsO_4. The secondary catalytic cycle is avoided.

The enantioselectivity of the ADH reaction obviously depends on the coordination properties of the chiral ligand L. Many such ligands have been screened. The most effective ones are those with chiral alkaloid units of the chincona family. The ligand 9.44 is one where a suitable spacer group couples two such units. These ligands coordinate to the OsO_4 molecule through the sp^3-hybridized nitrogen atom of the alkaloid unit. Although two alkaloid units are present in a ligand such as 9.44, coordination to only one OsO_4 molecule takes place. The presence of two alkaloid units increases the scope and enantioselectivity of the reaction.

(9.44)

9.4.1 Mechanism of ADH Reaction

It is clear from Fig. 9.10 that in the presence of L, enantioselection occurs during the conversion of OsO_4 to 9.41. Two possible paths, one involving (3+2) cycloaddition (Path A) and the other (2+2) cycloaddition (Path B) have been proposed. These are shown in Fig. 9.11. In the former, two new C–O bonds are formed simultaneously. In the latter first a C–O and an Os–C bond

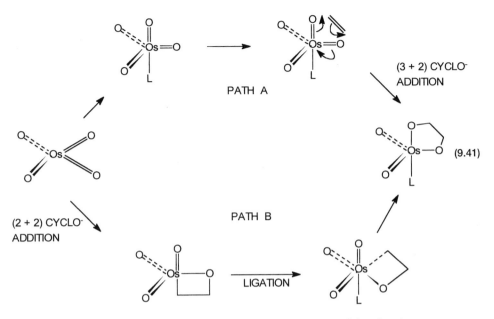

Figure 9.11   Two alternative pathways for the formation of the glycolate ester.

are formed. The Os–C bond then rearranges into another C–O bond. The other important difference between the two proposed mechanisms is that, in Path A ligation by L precedes (3+2) cycloaddition, while in Path B (2+2) cycloaddition is followed by ligation.

Evidence has been presented for both these mechanistic postulates. Such evidence has come from kinetic and structural data as well as isotope effect studies. Theoretical calculations aimed at explaining the observed kinetic isotope effects support the (3+2) cycloaddition mechanism. Two questions, however, remain to be answered. The mechanistic details of oxidation of an alkene by OsO_4 in the absence of L have not been studied in great detail. Also the changes, if any, that are effected by the introduction of L to such a mechanism are not known.

9.5 ASYMMETRIC CATALYTIC REACTIONS OF C–C BOND FORMATION

From what has been discussed so far in this chapter, it is clear that homogeneous catalysis has had spectacular success in imparting high enantioselectivities in the making of new C–H and C–O bonds. An enantioselective method for making new C–C bonds is also potentially very useful. Hydroformylation, hydrocyanation, and carbonylation are reactions that deal with the formation of new C–C bonds. All these have been turned into enantioselective catalytic systems with varying degrees of success. Considerable success has also been achieved in other C–C bond-forming reactions. In Section 9.2 for asymmetric cyclopropanation reaction the use of a copper complex with 9.8 as the ligand was mentioned. Several recent reports (see reference in answer to Question 24) indicate that this class of copper complexes has high potential as asymmetric catalysts for a number of other C–C bond-forming reactions. Such reactions include certain types of Diels–Alder and aldol condensation reactions. The important point is that in all these reactions the copper complexes function as chiral Lewis acids.

The alternative potential synthetic routes for the drug Naproxen® neatly illustrate the industrial significance of asymmetric hydroformylation and asymmetric hydrocyanation reactions. This is shown in Fig. 9.12. Regio- and enantioselective hydroformylation or hydrocyanation of 6-methoxy 2-vinyl naphthalene can give the desired enantiomers of the branched aldehyde or nitrile. These two intermediates can be oxidized or hydrolyzed to give S-Naproxen.

9.5.1 Asymmetric Hydroformylation Reaction

As shown by reaction 9.8, in an asymmetric hydroformylation reaction only the branched aldehyde product can have optical isomers. The linear aldehyde (shown within the brackets) is an undesirable byproduct. Successful asymmetric hydroformylation reaction thus needs to be chemo-, regio-, and enantioselective. Lack of chemoselectivity leads to hydrogenation rather than hydroformylation, and lack of regioselectivity results in the formation of the linear rather than the branched isomer.

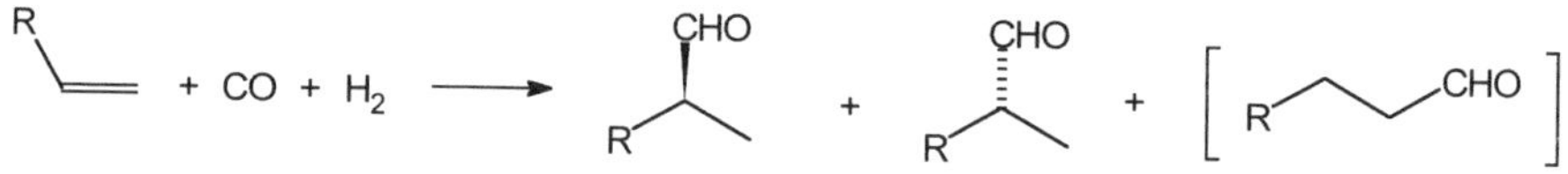

$$(9.8)$$

Hydroformylation of styrene and its analogues has attracted particular attention, since this provides a general method for the preparations of optically pure arylpropionic acids. Apart from Naproxen®, the drug ibuprofen is another arylpropionic acid–based nonsteroidal anti-inflammatory agent. As shown by 9.9, ibuprofen may in principle be synthesized by enantioselective hydroformylation reactions followed by oxidation of the aldehydic functionality.

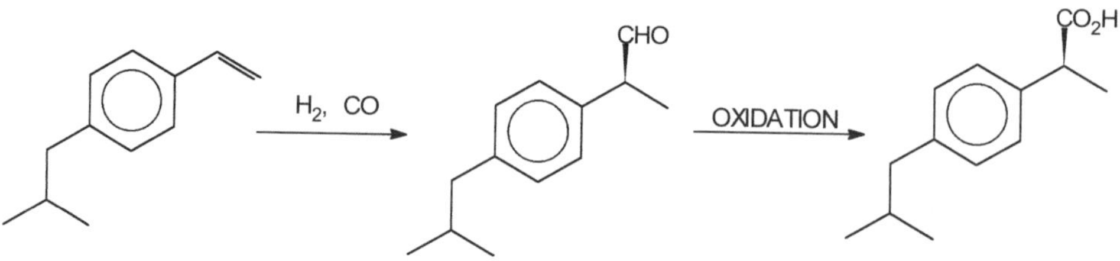

(S) Ibuprofen

$$(9.9)$$

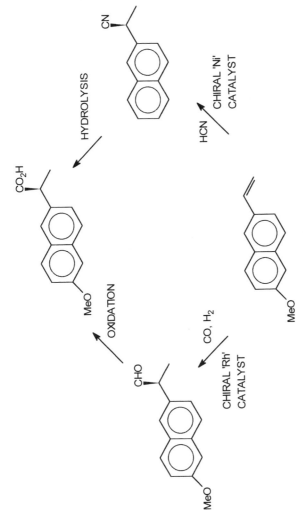

Figure 9.12 Two alternative synthetic routes for *S*-Naproxen. The left-hand route involves asymmetric hydroformylation, while the right-hand one uses asymmetric hydrocyanation.

The two metals that have been found to give encouraging conversions and selectivities for the hydroformylation of styrene are platinum and rhodium. The platinum-based catalytic system uses tin chloride as a promoter. It also uses triethyl orthoformate as a scavenger that reacts with the aldehyde to form the acetal. By removing it as soon as it is formed, any further degradative reactions of the aldehyde are avoided. The chirality in these reactions is induced by the use of optically active phosphorus ligands. With the best platinum catalyst, branched and linear aldehydes are produced in about equal proportion, but the former has an e.e. of >96%.

Until recently rhodium catalysts gave lower enantioselectivity, but higher chemoselectivity and activity, than the platinum-based catalysts. However, in the past few years rhodium complexes of a few chiral diphosphites and phosphinophosphito ligands have been reported. These complexes have excellent activities and high chemo-, regio-, and enantioselectivities.

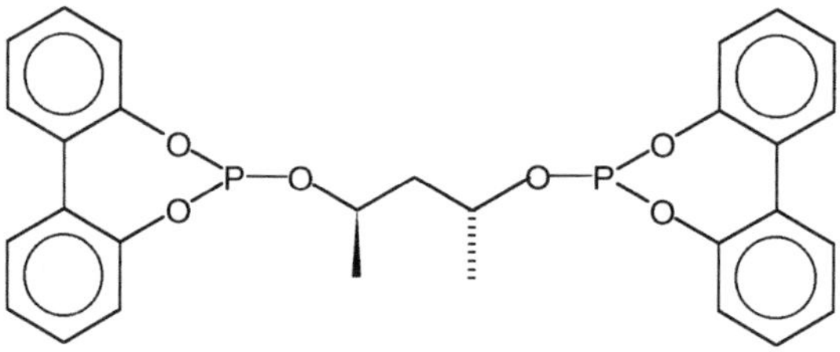

(9.45)

The phosphite ligand 9.45 is derived from the chiral back-bone of 2,4-pentane diol while 9.46 is based on binapthol and is called (R,S) BINAPHOS. With the latter, under mild conditions high regioselectivity (>85%) and enantioselectivity (>95%) are obtained for a wide variety of styrene derivatives including the precursor for ibuprofen.

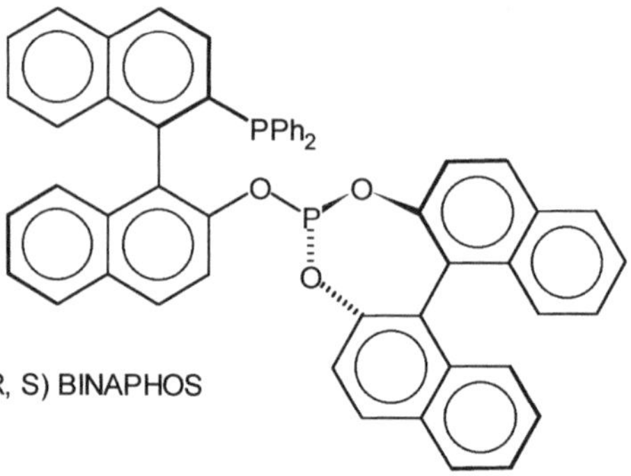

(R, S) BINAPHOS

(9.46)

9.5.2   Mechanism of Asymmetric Hydroformylation Reaction

As we have seen in Chapter 5, the mechanistic details of the hydroformylation reaction with rhodium triphenylphosphine complexes are well established. These mechanistic considerations may be modified and extrapolated to the chiral hydroformylation system. One important point to bear in mind is that bidentate rather than monodentate ligands are involved in the chiral hydroformylation system.

A hypothetical catalytic cycle for asymmetric hydroformylation reaction is shown in Fig. 9.13. The precatalyst Rh(acac)(P-P) reacts with H_2 and CO to give the square planar catalytic intermediate 9.47. Alkene addition to 9.47 can lead to the formation of 9.48, 9.49, and 9.50. The steric requirements of the chelating ligand would have to be such that the formation of 9.50 is avoided. This is because alkene insertion into the Rh–H bond in this case would lead to the formation of the linear rather than the branched alkyl. Both 9.48 and 9.49, which differ in the coordination positions of the phosphorus atoms, can give 9.51, which has the desired branched alkyl ligand.

Both regioselection and enantioselection are effected during the conversion of 9.48 or 9.49 to 9.51. The alkyl complex 9.51 undergoes the usual CO insertion followed by oxidative addition of dihydrogen, etc. (see Section 5.2.1) to give enantiomerically pure aldehyde product. Although in the catalytic cycle only one diastereomer for 9.48 and 9.49 has been shown, both these intermediates can have two diastereomers each. In these diastereomers the chirality of the chelating ligand is obviously the same, but coordination by the alkene takes place through the other enantioface.

Two alternative postulates for enantioselection may be proposed. In the first all four diastereomers are formed in varying amounts, and the relative amounts are determined by their respective thermodynamic stabilities. Assuming approximately equal amounts of each diastereomer to be present, if one of them has a transition state that is about 2.5 kcal lower in energy than the transition states of the others, more than 90% enantioselection for 9.51 would result. The proposed mechanism for enantioselection is thus similar to that of asymmetric hydrogenation (see Section 9.3.1). In the second postulate only one such diastereomer is produced; that is, the thermodynamic stability of one of the diastereomers is higher than that of the others. The stable diastereomer owing to its steric and electronic characteristics is converted to 9.51 in an enantioselective manner.

Either of the mechanisms postulated above is far from being well established. With 9.46 as a ligand, in the absence of alkene but in the presence of H_2 and CO, a species such as 9.52 has been identified by NMR. This complex is sufficiently stable. No fluxional behavior is observed up to a temperature of 60°C.

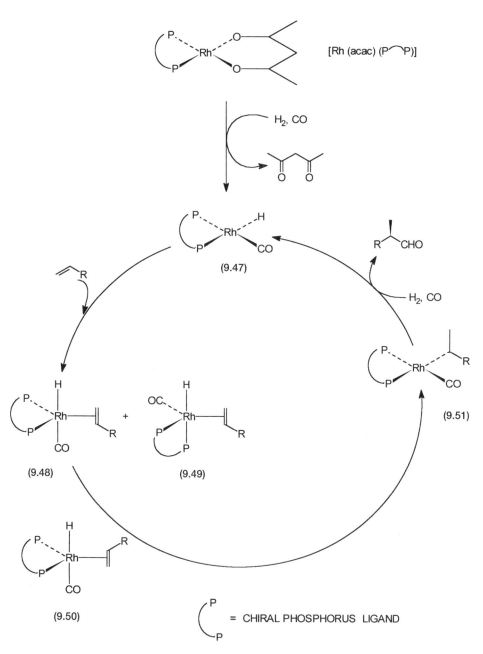

[Rh (acac) (P⌒P)]

(9.47)

(9.48)

(9.49)

(9.50)

(9.51)

⌒(P–P) = CHIRAL PHOSPHORUS LIGAND

Figure 9.13 Proposed catalytic cycle for asymmetric hydroformylation reaction with [Rh(acac)(P–P)] as the precatalyst. The chelating chiral phosphorus ligand could be 9.45 or 9.46. acac = acetylacetonato anion.

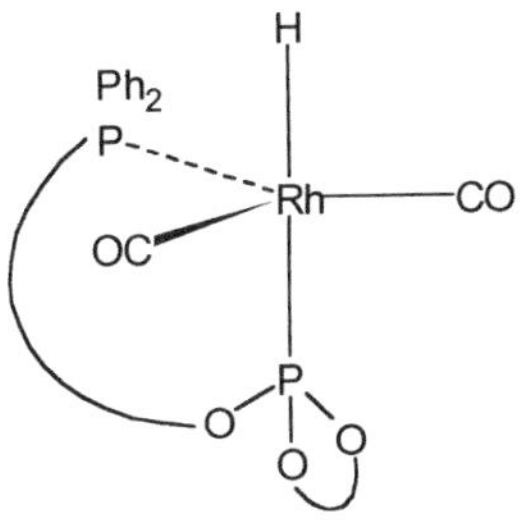

(9.52)

Obviously the preferred coordination positions are the phosphite in the axial position and the phosphine in the equatorial position. This, however, is not the case with 9.45. Here NMR studies indicate the formation of a relatively stable isomer, where both the phosphorus atoms are in the equatorial positions as in 9.48. The evidence gathered so far seems to indicate that for both 9.45 and 9.46 only one stable diastereomer is formed. As proposed in the second postulate, this diastereomer probably determines the stereochemical course of the subsequent catalytic steps.

9.5.3 Asymmetric Hydrocyanation Reaction

Successful development of the asymmetric hydrocyanation reaction may provide a versatile route to chiral nitriles, amines, and acids. As we have seen, the mechanistic details of the hydrocyanation reaction of butadiene with zerovalent nickel complexes are well established. By using a nickel complex of a chiral bidentate phosphinite ligand, 9.53, good conversion and enantioselectivity (>85% e.e.) for the hydrocyanation of 6-methoxy 2-vinyl naphthalene have been obtained.

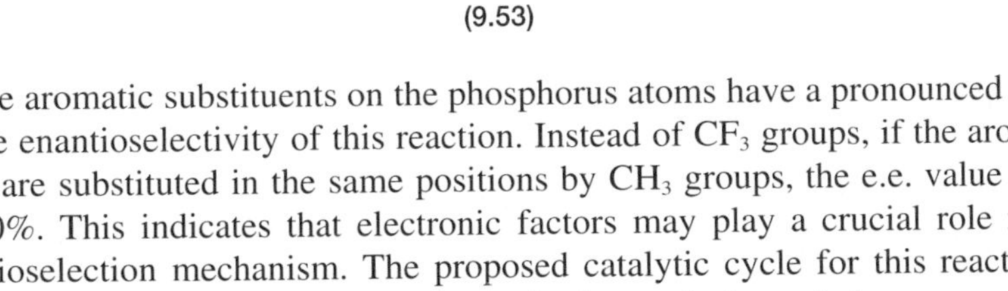

(9.53)

The aromatic substituents on the phosphorus atoms have a pronounced effect on the enantioselectivity of this reaction. Instead of CF_3 groups, if the aromatic rings are substituted in the same positions by CH_3 groups, the e.e. value drops by 70%. This indicates that electronic factors may play a crucial role in the enantioselection mechanism. The proposed catalytic cycle for this reaction is shown in Fig. 9.14. All the steps shown in the catalytic cycle have precedence in achiral hydrocyanation reactions (see Section 7.7).

The precatalyst $Ni(COD)_2$ reacts with 9.53 to give 9.54. This intermediate undergoes substitution of the remaining COD molecule by the substrate to give

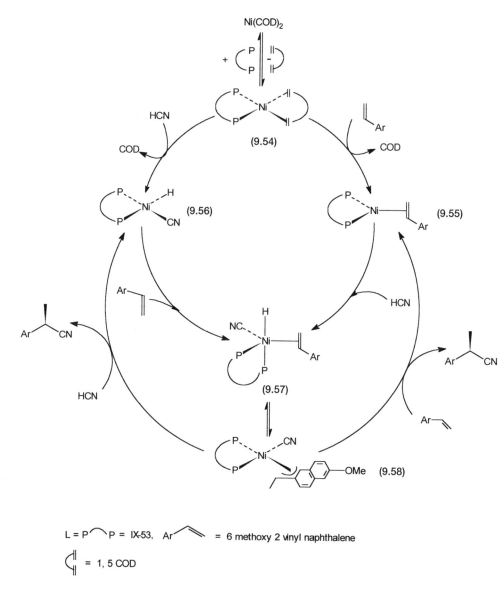

L = P⌒P = IX-53,   Ar⌒⧹⧹ = 6 methoxy 2 vinyl naphthalene

⧸⧹ = 1, 5 COD

Figure 9.14   Catalytic cycle for the asymmetric hydrocyanation of 6-methoxy-2-vinyl naphthalene with Ni(COD)$_2$ as the precatalyst and 9.53 as the chiral ligand.

9.55. Oxidative addition of HCN onto 9.54 with the elimination of COD leads to the formation of 9.56. Oxidative addition by HCN and coordination by the substrate onto 9.55 and 9.56, respectively, lead to the formation of 9.57. Insertion of the alkene functionality into the Ni–H bond leads to the formation of the η^3-allyl intermediate 9.58. Substrate addition or oxidative addition of

HCN may accompany reductive elimination of the nitrile product from 9.58. In the former case 9.55 and in the latter 9.56 are regenerated to complete the catalytic cycle.

Kinetic analyses indicate that both the right-hand and the left-hand-side loops of the catalytic cycle are active. Of the two, which dominates would depend upon the relative concentrations of the substrate and HCN. Kinetic and deuterium-labeling studies also indicate all the steps from 9.54 to 9.57 to be reversible. The intermediate 9.55 has been identified by ^{31}P NMR. Note that for 9.55 more than one diastereomer is possible, and indeed at room temperature these are found to be in rapid equilibrium with each other. Kinetic and spectroscopic evidence taken together suggests that enantioselectivity is determined in the insertion, 9.57 to 9.58, and/or in the reductive elimination, 9.58 to 9.55 or 9.56, steps.

9.5.4 Nitroaldol Condensation

Asymmetric nitroaldol condensation, reaction 9.10, has been used to make an intermediate for the drug S-propanolol 9.61. The condensation reaction is catalyzed by the chiral heterobimetallic binaphthol complex 9.12 with high (>90%) enantioselectivity.

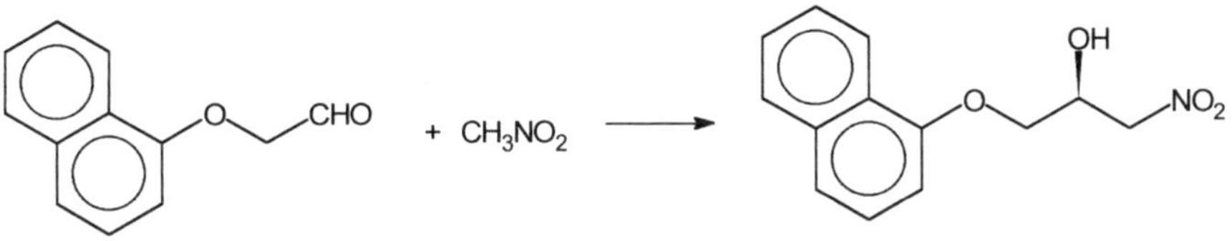

(9.10)

Complex 9.12 and its analogues, with other rare earth and alkali metals, have the unique combination of properties of both a Brønsted base and a Lewis acid. The Li–O bond (Brønsted base) abstracts a proton from nitromethane. At the same time coordination of the carbonyl oxygen atom to La^{3+} (Lewis acid) activates the carbonyl group. The formulations for 9.12 and its analogues were first established by a sophisticated mass-spectrometric technique called laser-desorption/ionization time-of-flight (LDI-TOF) mass spectrometry. The molecular structures of several such complexes have subsequently been established by single-crystal X-ray studies.

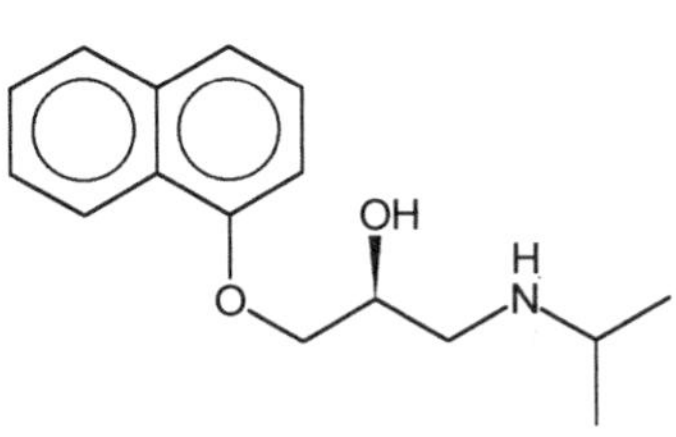

S-PROPANOLOL

(9.61)

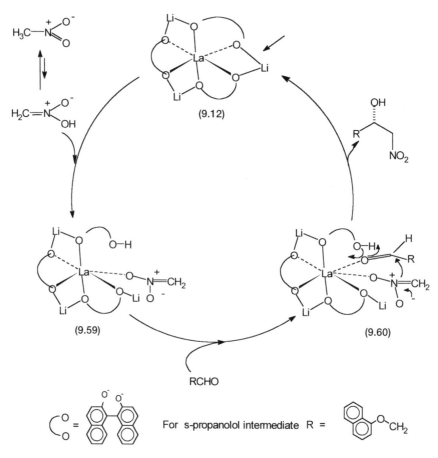

Figure 9.15 Catalytic cycle for asymmetric nitroaldol condensation reaction with 9.12 as the chiral catalyst. The La—O bond marked by an arrow opens up due to protonation of the O atom by nitromethane.

The proposed catalytic cycle for the nitroaldol condensation reaction is shown in Fig. 9.15. The Li—O bond marked by an arrow in 9.12 is cleaved due to protonation of the binaphthol oxygen atom by nitromethane. In 9.59 one of the oxygen atoms of the nitro group coordinates to the La^{3+} cation. Conversion of 9.12 to 9.59 is facilitated by the Brønsted base behavior of the Li—O bond and the Lewis acidity of La^{3+}. The activated CH_2NO_2 moiety thus produced attacks the C atom of the carbonyl group. This process, as shown by curly arrows in 9.60, is facilitated by an interaction between the carbonyl oxygen atom and La^{3+}. Elimination of nitroalcohol takes 9.60 back to 9.12, completing the catalytic cycle. The chiral binaphthol ligands probably ensure that coordination to La^{3+} by R′CHO is enantioface selective.

PROBLEMS

1. Give the absolute descriptors (i.e., R and S assignments) to the chiral C atoms in 9.7, 9.8, 9.9, 9.10, 9.15, and 9.18. Draw the structures of methyl-(Z)- and methyl-(E)-acetamido cinamates.

Ans. See chapters on stereochemistry in the books by T. W. G. Solomons or R. T. Morrison and R. N. Boyd. R. Noyori, *Chemtech*, June, 361, 1992, has a one-page description of the stereochemical nomenclatures and rules.

2. (a) Why are the ligands shown in Fig. 9.1 not used for rhodium-based hydrogenation reactions and (b) ligands of Fig. 9.2 not used for epoxidation reactions?

Ans. (a) Rhodium catalysts for hydrogenation require ligands capable of stabilizing a low oxidation state. (b) Oxygen atom donors oxidize phosphine to phosphine oxide. Also phosphines are not the best ligands to stabilize Ti^{4+}, Mo^{6+}, etc.

3. What metal and ligand combinations are used for the asymmetric hydrogenation of (a) imines, (b) ketones, (c) alkenes with an α-acetamido group? What is the combination for cyclopropanation?

Ans. (a) Ir, 9.15, (b) 9.16, (c) 9.13, Rh or 9.14. Cu and 9.8.

4. With R,R DIPAMP as the ligand and methyl-(Z)-α-acetamido cinamate as the substrate, 9.19 and 9.20 are the major and minor diastereomers, respectively. Let 9.19' and 9.20' be the analogues of 9.19 and 9.20, with S,S DIPAMP as the ligand. Predict the order of the rate constants for the oxidative addition of H_2 in the following cases: (a) 9.19' and 9.20', (b) 9.19 and 9.20', (c) 9.19' and 9.20, (d) 9.19 and 9.19'.

Ans. Let the four rate constants be k_1, k_2, k_1', and k_2'. Note that 9.19 and 9.20' have an enantiomeric relationship as do 9.19' and 9.20. It is known that $k_2 > k_1$. (a) $k_1' > k_2'$, (b) $k_1 = k_2'$, (c) $k_1' = k_2$, (d) $k_1 < k_1'$.

5. Assuming the free energy values shown in Fig. 9.5 are measured at 20°C, calculate the concentration ratio of 9.19 to 9.20 and the ratio of the rate constants for the oxidative addition of dihydrogen.

Ans. From $-RT \ln K = \Delta G°$, $K \sim 10 (C_{major}/C_{minor})$. From $\ln(k_1/k_2) = \Delta\Delta G^{\neq}/RT$, $\log(k_1/k_2) = 2.77(\Delta\Delta G^{\neq} = 2.3 + 1.4 = 3.7$ kcal mole$^{-1})$.

6. In Figs. 9.3 and 9.6 the precatalyst 9.14 is converted to 9.19, 9.20, and 9.32. What is the intermediate involved?

Ans. 9.28.

7. A racemic mixture of methyl 2-(α-hydroxybenzyl) acrylate [$MeO_2C-C(:CH_2)CH(OH)Ph$] is hydrogenated with 9.14 (chiral ligand = R,R DIPAMP) as the precatalyst up to about 50% conversion. The starting

material recovered from the reaction shows optical activity. Explain this observation. How many stereoisomers are expected in this reaction?

Ans. Optical activity is due to kinetic resolution. Four stereoisomers of the hydrogenated product are expected (see J. M. Brown, *Angew. Chem. Int. Ed. Engl.* **26**, 190–203, 1987).

8. In Fig. 9.6, conversion of 9.32 to 9.33 involves coordination by allyl amine followed by hydride abstraction. Modify the catalytic cycle to show this additional step, as well as conversion of 9.14 to 9.28 to 9.32.

Ans. See R. Noyori in *Chem. Soc. Rev.* **18**, 187–208, 1989.

9. What are the similarities and differences in the behavior of α-acetamido-cinamic acid and allyl amine as ligands in asymmetric hydrogenation and isomerization reactions, respectively?

Ans. Similarity: Both prochiral and undergo enantioface selective alkene insertion into M–H bond (hydride attack). Difference: The first acts as a chelating ligand all throughout the catalytic cycle, but the second only in 9.32.

10. In Fig. 9.7 conversions of 9.36 to 9.37 must involve another intermediate. What is it?

Ans. Epoxy allyl alkoxide and a solvent molecule coordinated to the Ti^{4+} ion. The former is then protonated and displaced by allyl alcohol.

11. The room-temperature NMR spectrum of 9.35 with a bulky R group has two signals for the CO_2R groups. When the R group is small, only one signal is observed at room temperature, but two signals are observed at low temperatures. Explain.

Ans. The molecules with small R are fluxional. At low temperature the rate of interchange between bridging and free CO_2R groups is slower than the NMR time scale.

12. At $-20°C$, Ti-tartarate plus the hydroperoxide-based catalytic system give near-identical enantioselectivities for a variety of allyl alcohols with different substituents on C2 and C3. What could be concluded from this observation?

Ans. As different substituents have little or no effect, stereoelectronic rather than only steric factors are responsible for enantioselection.

13. Asymmetric epoxidation of unfunctionalized alkenes by NaOCl with 9.38B-type catalysts is found to be substantially accelerated in the presence of near-catalytic quantities of amine N-oxides. What is the mechanistic significance of this observation?

Ans. Probable axial coordination by N-oxide (in place of Cl^-) to stabilize 9.39 (see N. S. Finney et al., *Angew. Chem. Int. Ed. Engl.* **36**, 1720–23, 1997).

14. In Fig. 9.11 assign oxidation states to all the osmium-containing species and identify the redox steps. Is it possible to form 9.42 and 9.43 in an organic solvent?

Ans. 9.41, 6+; 9.42, 6+; 9.43, 8+; the redox steps are OsO_4 to 9.41 and 9.42 to 9.43. No; formation of 9.42 and 9.43 require HO^- and $[Fe(CN)_6]^{3-}$, both of which as alkali metal salts are insoluble in organic solvents.

15. Assuming that the organic ligand remains attached to the metal, what may be a possible oxidation product of 9.41, and under what conditions may this be expected to be formed?

Ans. OsO_3(glycolate); in an organic solvent, where hydrolysis is slow and with an organic-solvent-soluble oxidant such as NMO.

16. In what way does the presence of two alkaloid units in 9.44 help in improving the enantioselectivity? Why is asymmetric ADH reaction considered to be a very important reaction in synthetic organic chemistry?

Ans. The spatial orientation of the aromatic part of the second alkaloid unit creates "pocket" where an enantioselective fit of the substrate occurs (see H. C. Kolb et al., *Chem. Rev.* **94**, 2483–547, 1994; E. J. Corey et al., *J. Am. Chem. Soc.* **118**, 11038–53, 1996). In the ADH reaction *syn*-addition across double bond and generation of two stereocenters in an enantioselective manner are achieved in one step.

17. Osmium tetroxide on treatment with the chelating diamine L_2 (L_2 = [R,R]-*trans*-1,2 bis[N-pyrrolidino]cyclohexane) gives OsO_4L_2. What is the electron count in this complex, and does this have any relevance to the ADH mechanism?

Ans. 20. Indirect evidence for mechanism as outlined by Path A in Fig. 9.11 (see E. J. Corey et al., *J. Am. Chem. Soc.* **118**, 7851–52, 1996).

18. The catalytic system based on the precatalyst [Rh(acac)(*R,S* BINAPHOS)] shows much higher activity for asymmetric hydroformylation of styrene than that based on $RhH(CO)L_3$ (L=PPh_3) plus *R,S* BINAPHOS. Why?

Ans. Formation of inactive intermediates such as HRh(CO)*L*(BINAPHOS) in equilibrium with 9.49.

19. Draw structures of ligands derived from the chiral framework of glucose, tartaric acid, binaphthol, and cinchona alkaloids that are used for efficient asymmetric hydrocyanation, epoxidation, hydroformylation, and alkene dihydroxylation reactions respectively.

Ans. 9.53, 9.10, 9.46, and 9.44, respectively.

20. $NiLBr_2$ (L = 9.53) is reacted with 6-methoxy-2-vinyl naphthalene and zinc powder to give a substance ^{31}P NMR spectrum (^{1}H decoupled), of which at $-30°C$ shows eight doublets. These coalesce into a broad singlet at

70°C. What is material, and what is the significance of the NMR data in the asymmetric hydrocyanation reaction?

Ans. The material is 9.55. The substrate has two enantiofaces and each enantioface can have either up or down orientation with respect to NiL, which has C_1 symmetry. This gives four diastereomers with two P atoms, each being magnetically inequivalent. At higher temperatures the diastereomers interconvert. 9.54 to 9.55 cannot be the enantioselection step (see A. L. Casalnuovo et al., *J. Am. Chem. Soc.* **116**, 9869–82, 1994).

21. The e.e. of the nitroaldol condensation reaction between PhCHO and CH_3NO_2 catalyzed by 9.12-type complexes with different rare earth metals are: <20% for Yb^{3+}, 40% for Y^{3+}, and >70% for Eu^{3+},Sm^{3+}. Give a plausible explanation.

Ans. e.e. correlated with ionic radii of the central metal ion (see M. Shibasaki et al., *Angew. Chem. Int. Ed. Engl.* **36**, 1237–56, 1997).

22. Chiral amino alcohols catalyze the enantioselective reaction of diethyl zinc with benzaldehyde to give 1-phenyl propanol. A mixture of the two enantiomers of the amino alcohol, *not* in equal proportion, is used as the catalyst, and the relative amounts varied. More than 90% e.e. is obtained with a ratio of 1.2, but zero e.e. is obtained when it is 1.0 (exactly racemic). Explain.

Ans. Nonlinear effect due to the presence of several diastereomers (see M. Kitamura et al., *J. Am. Chem. Soc.* **120**, 9800–9, 1998).

23. (a) In metal-catalyzed asymmetric cyclopropanation, what may be a possible mechanism for enantioselection? (b) Treatment of $[CuL]^+$ (L = bis-azaferrocene), an active cyclopropanation catalyst, with styrene gives crystals of $[CuL(styrene)]^+$. What mechanistic insight may this structure provide?

Ans. (a) Formation of diastereomeric metallocyclobutanes as intermediates and/or transition states of different energies. (b) Whether or not the enantioface of styrene in the structure undergoes carbene addition can be determined (see M. M. C. Lo and G. C. Fu, *J. Am. Chem. Soc.* **120**, 10270–71, 1998).

24. Among many novel chiral phosphines, dendrimers containing chiral diphosphines may be of special practical importance. Why?

Ans. Ease of catalyst separation by membrane filtration or precipitation (see C. Kollner et al., *J. Am. Chem. Soc.* **120**, 10274–75, 1998).

25. Equimolar quantities of $CuCl_2$ and (*S,S*)-bis(phenyloxazolinyl) pyridine (an analogue of 9.8 with pyridine substituted at the 2,6 positions by chiral oxazoline units) were reacted with excess $AgSbF_6$ to give a blue solution. On addition of $BnOCH_2CHO$ (Bn = benzyl) to the solution followed by slow evaporation, deep blue crystals were obtained. Propose a formula and a structure for the crystals. What is the relevance of this structure to asymmetric catalysis?

Ans. Five-coordinated square-pyramidal [CuL(BnOCH₂CHO)] [L = (*S,S*)-bis (phenyloxazolinyl) pyridine]. A model catalytic intermediate in Cu−L-based asymmetric aldol condensation reactions (see D. A. Evans et al., *J. Am. Chem. Soc.* **121**, 669−85; 7559−73 and 7582−94, 1999).

BIBLIOGRAPHY

For all the sections

Books

Chapter on stereochemistry in *Organic Chemistry*, by T. W. G. Solomons, Wiley, New York, 5th edition, 1992.

Chapter on stereochemistry in *Organic Chemistry*, by R. T. Morrison and R. N. Boyd, Allyn and Bacon, Boston, 5th edition, 1987.

Asymmetric Catalysis in Organic Synthesis, by R. Noyori, Wiley, New York, 1994.

Catalytic Asymmetric Synthesis, ed. by I. Ojima, VCH, New York, 1993.

Asymmetric Synthesis, ed. by J. D. Morrison, Academic Press, New York, Vol. 5, 1985.

Discussion on some asymmetric homogeneous catalytic reactions can also be found in *Homogeneous Catalysis: The Applications and Chemistry of Catalysis by Soluble Transition Metal Complexes*, by G. W. Parshall and S. D. Ittel, Wiley, New York 1992.

Inorganic Chemistry, by D. Shriver, P. Atkins, and C. H. Langford, W. H. Freeman, New York, 1994.

Advanced Inorganic Chemistry, by F. A. Cotton and G. Wilkinson, Wiley, New York, 1988.

Principles and Applications of Organotransition Metal Chemistry, by J. P. Collman and L. S. Hegedus, University Science Books, Mill Valley, California, 1987.

The Organometallic Chemistry of the Transition Metals, by R. H. Crabtree, Wiley, New York, 1994.

Also see Section 2.9 of Vol. 1 of *Applied Homogeneous Catalysis with Organometallic Compounds*, ed. by B. Cornils and W. A. Herrmann, VCH, Weinheim, New York, 1996.

Articles

Sections 9.1 and 9.2

For recent studies on thalidomide, see G. W. Muller, *Chemtech* **27**(1), 21−25, 1997.

For general discussion on chirality and asymmetric catalysis, see W. A. Nugent et al., *Science* **259**(5094), 479−83, 1993.

For chiral drugs, see S. C. Stinson, *Chem. Eng. News*, **75**(42), 38−70, 1997, and **76**(39), 83−104, 1998.

For compounds lised in Table 9.1, see R. Noyori and S. Hasiguchi, in *Applied Homogeneous Catalysis with Organometallic Compounds* (see under books), Vol. 1, pp. 552–73; H. Blaser and F. Spindler, *Chimia* **51**, 297–99, 1997; R. A. Sheldon, *Chimia* **50**, 418–19, 1996; R. Schmid, *Chimia* **50**, 110–13, 1996.

For ligands in Figs. 9.1 and 9.2, see: D. A. Evans et al., *J. Am. Chem. Soc.* **113**, 726–28, 1991, 9.8; E. N. Jacobsen et al., *J. Am. Chem. Soc.* **113**, 7063–64, 1991, 9.9; M. G. Finn and K. B. Sharpless, in *Asymmetric Synthesis*, ed. by J. D. Morrison, Academic Press, New York, Vol. 5, 1985, pp. 247–308, 9.10; R. Noyori and H. Takaya, *Acc. Chem. Res.* **23**, 345–50, 1990, 9.11 and 9.16; M. Shibasaki et al., *Angew. Chem. Int. Ed.* **36**, 1237–56, 1997, 9.12; H. B. Kagan, in *Asymmetric Synthesis*, ed. by J. D. Morrison, Academic Press, New York, Vol. 5, 1985, pp. 1–40, 9.13 and other chiral phosphines; W. S. Knowles, *J. Chem. Edu.* **63**, 222–25, 1986, 9.14; H. Blaser and F. Spindler, *Chimia* **51**, 297–99, 1997, 9.15; R. Schmid, *Chimia* **50**, 110–13, 1996, 9.17; M. J. Burk, *J. Am. Chem. Soc.* **113**, 8518–19, 1991, 9.18.

Section 9.3

For asymmetric hydrogenation, see J. Halpern, in *Asymmetric Synthesis*, ed. by J. D. Morrison, Academic Press, New York, Vol. 5, 1985, pp. 41–70.

J. M. Brown and P. A. Chaloner, in *Homogeneous Catalysis with Metal Phosphine Complexes*, ed. by L. H. Pignolet, Plenum Press, New York, 1983, pp. 137–65.

For 'binap'-based asymmetric hydrogenation and isomerization, see R. Noyori, *Chem. Soc. Rev.* **18**, 187–208, 1989; S. Otsuka and K. Tani, in *Asymmetric Synthesis*, ed. by J. D. Morrison, Academic Press, New York, Vol. 5, 1985, pp. 171–93.

For asymmetric epoxidation of allyl alcohols, see M. G. Finn and K. B. Sharpless, in *Asymmetric Synthesis*, ed. by J. D. Morrison, Academic Press, New York, Vol. 5, 1985, pp. 247–308.

For epoxidation of unfunctionalized alkenes, see M. Palucki et al., *J. Am. Chem. Soc.* **120**, 948–54, 1998.

For asymmetric hydrolysis, see M. Tokunaga et al., *Science* **277**(5328), 936–38, 1997.

Section 9.4

See references given in answer to Problem 16. Also see A. M. Rouhi in *Chem. Eng. News* **75**(44), Nov. 3, 23–26, 1997.

Section 9.5

For asymmetric hydroformylation, see S. Gladiali et al., *Tetrahedron-Asymmetry* **6**, 1453–74, 1995; F. Agbossou et al. *Chem. Rev.* **95**, 2485–2506, 1995.

For asymmetric hydrocyanation, see reference given in answer to Problem 20.

For asymmetric nitroaldol reaction, see reference given in answer to Problem 21.

INDEX

Page references in bold signify that a definition or exhaustive discussion on the topic will be found on that particular page.